PRACTICAL GUIDES TO TESTING AND COMMISSIONING OF MECHANICAL, ELECTRICAL AND PLUMBING (MEP) INSTALLATIONS

CHANDRA B. GURUNG

PARTRIDGE

Copyright © 2018 by Chandra B. Gurung.

ISBN: Softcover   978-1-5437-4691-4
        eBook   978-1-5437-4690-7

All rights reserved. No part of this book may be used or reproduced by any means, graphic, electronic, or mechanical, including photocopying, recording, taping or by any information storage retrieval system without the written permission of the author except in the case of brief quotations embodied in critical articles and reviews.

Because of the dynamic nature of the Internet, any web addresses or links contained in this book may have changed since publication and may no longer be valid. The views expressed in this work are solely those of the author and do not necessarily reflect the views of the publisher, and the publisher hereby disclaims any responsibility for them.

Print information available on the last page.

To order additional copies of this book, contact
Toll Free 800 101 2657 (Singapore)
Toll Free 1 800 81 7340 (Malaysia)
orders.singapore@partridgepublishing.com

www.partridgepublishing.com/singapore

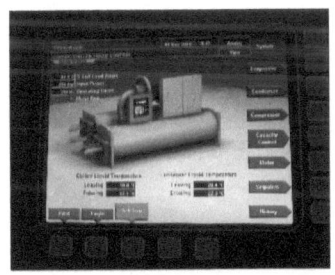

Practical Guides to Testing and Commissioning of Mechanical, Electrical and Plumbing (MEP) Installations

By Captain (Retired) Chandra B. Gurung, QGE, MCQI, MIET

PREFACE

During my carrier as a Commissioning Manager with Commtech Asia in Hong Kong and other parts of Asia Pacific I came across with many MEP Contractors. Some were good in preparing their T&C Method Statements and some had difficulties. The contract documents make reference to British Standards, EN Standards, Local Code of Practices, ASHRAE and CIBSE Standard & Codes. But it does not tell you how to prepare a very simple T&C method Statement.

In late 70s I used to work with Architectural Services Department of Hong Kong SAR Government, as a Building Services Technical Officer for 5 years on Secondment from British Army and their T&C Procedure was found to be very useful. This helped me a lot in doing my T&C works later in Tsing Ma Bridge and Ting Kau Bridge projects in Hong Kong with Highways Department where I was employed as a Resident Site Staff.

This book will provide guide lines for Electrical Engineers, Mechanical Engineers and Fire Services Engineers on how to prepare technical parts of a T&C Method Statement submission for their MEP contracts. For Project Directors, Project Managers and Resident Staff it serves as a check list to ensure that all equipment are tested properly for energy saving and their resilience.

In late 90s before the millennium I was working in MCI World Data Centre in Chai Wan, Hong Kong, when I came across with many critical electrical equipment like UPS, PDU, STS, CRAC and Gensets. This helped me a lot in 2007 when I started working with Commtech Asia in Hong Kong as a T&C Manager. From time to time the Architectural Services Department keep updating their T&C Procedure and this has always been my source of information in writing this book and sincerely would like to thank them.

It is to be noted that in all T&C process each procedure requires Test Sheet to record the test results, due to limited space on this book I am unable to provide all sample of these test sheets to readers for their reference.

At the end of this book I have appended two appendixes. Appendix A gives readers some guide lines on how to prepare an Integrated System Test Method Statement with example of 3 failure test scenarios. In order to prepare this Method Statement one, have to study all electrical schematic diagrams carefully. Then analyzing all circuits fed by that transformer i.e. if Tx No. 5 fails which part of the building will lose Electrical power in stage 1 (Before Genset starts) and which part of the building will resume power in stage 2 (after the Genset has started). If there is an UPS unit in the circuit during this transition period UPS will switch to Battery Mode to take over the UPS Load, then back to Inverter Mode when Genset ramps up the power supply as a backup source.

Appendix B is a sample of Heat Load Test Method statement. The main purpose of a Heat Load Test is to test the failure scenarios related to change-over and operation of Chillers, Cooling Towers and related Chilled water and Condenser pumps.

As it can be seen the load on the Data Centre is slowly increase in steps from 0 – 100% Load known as staging up ensuring that the room temperature and humidity is fully stabilized prior to any change-over action for next stage. For these two Appendixes I would like to thank Commtech Asia Limited for their consent in allowing me to use these sample Method Statements.

DISCLAIMER

This Practical Guides to Testing and Commissioning of MEP Installations lay down general guide lines for technical section of a T&C Method Statement for nearly all MEP services encountered in Construction Industry. This book will provide Guide Lines for Electrical Engineers, Mechanical Engineers, Plumbing and Drainage Engineers and Fire Services Engineers to prepare their Technical Procedure part of the T&C Method Statement. However, it does not cover full section of T&C Method Statement like Introduction, Tools and Instruments Requires, Resources, Materials, Storage and Handling, Responsibilities and Safety & Risk Assessment etc.

Users who choose to adopt this Practical Guides for their works are responsible for making their own assessments and judgement of all information contained herein. The Author does not accept any liability and responsibility for any special, indirect or consequential loss or damage whatsoever arising out of or in connection with the use of this Practical Guides or reliance placed on it.

ABBREVIATION

AC = Alternating Current
A/C = Air conditioning
ACB = Air Circuit Breaker
AFA = Automatic Fire Alarm
AHU = Air Handling Unit
ATS = Auto-Transfer Switch
BMS = Building Management System
CCS = Chiller Control System
CCTV = Closed Circuit Television
CDWP = Condensing Water Pump
CHWP = Chilled Water Pump
CHWS = Chilled Water Supply
CHWR = Chilled Water Return
CPC = Circuit Protective Conductor
COP = Code of Practice
CRAC = Computer Room Air Conditioning
CT = Current Transformer
CVM = Current Volt Meter
DC = Direct Current
DOL = Direct On Line
DRUPS = Dynamic Rotary UPS
DRV = Double Regulating Valve
EAF = Exhaust Air Fan
E/F = Earth Fault
ELV = Extra Low Voltage
ES = Earthing Switch
E-Stop = Emergency Stop
FAD = Fresh Air Duct
FAF = Fresh Air Fan
FAT = Factory Acceptance Test
FCU = Fan Coil Unit

FLC = Full Load Current
F.S. = Fire Services
Genset = Generator Set
HV = High Voltage
HLT = Heat Load Test
HVPF = High Voltage Power Frequency
Hz = Hertz
IP = Independent Parallel
IST = Integrated System Test
kV = Kilo Volt
kVar = Kilo Volt Amps Reactive
Kw= Kilo Watt
Kwh = Kilo Watt Hour
L/min = Litre Per minutes
LMCP = Local Motor Control Panel
LV = Low Voltage
LVSB = Low Voltage Switch Boards
M = Mega
MCB = Miniature Circuit Breaker
MCC = Motor Control Centre
MEP = Mechanical Electrical and Plumbing
MV = Medium Voltage
NER = Neutral Earth Resistor
O/C = Over Current
PAU = Pressurized Air Unit
PDU = Power Distribution Unit
P.F. = Power Factor
PSI = Per Square Inch
PQM = Power Quality Meter
RCD = Residual Current Device
REW = Registered Electrical Worker
RH = Relative Humidity
Rpm = Revolutions per Minutes
SAT = Site Acceptance Test
SCADA = Supervisory Control and Data Acquisition

STS = Static Transfer Switch
T&C = Testing and Commissioning
THD = Total Harmonics Distortion
Tx = Transformer
UPS = Uninterruptible Power Supply
VAV = Variable Air Volume
VCB = Vacuum Circuit Breaker
VESDA = Very Early Smoke Detection Alarm
VFD = Variable Frequency Drive
V min = V Minimum
V max = V Maximum
WLD = Water Leakage Detection System
Y-D = Star Delta

CONTENTS

Chapter 1 Introduction 1
Chapter 2 Electrical System 2
Chapter 3 Mechanical Ventilation and Air-conditioning System 27
Chapter 4 Fire Services System 59
Chapter 5 Plumbing and Drainage System 70
Chapter 6 Building Management System 80
Chapter 7 Appendices 92

Appendix A – IST Method Statement Sample 93
Appendix B – HLT Method Statement Sample 113

CHAPTER 1

Introduction

In old days for all Government Projects T&C would have been carried out by Residence Site Staff Inspectors and by MEP consultants for Private Buildings. Nowadays in large projects it is quite common to employ an Independent T&C Consultant to carry out such works. This way the MEP Consultant can concentrate on other expects of the projects like design, drawing submission, material submission, installation checking and other aspect of installation progress. And the T&C consultant can concentrate on T&C method statement from the contractors and carry out detailed T&C works for the project leading to Heat Load Test and Integrated System Test.

After the completion of all Testing and Commissioning activities an Integrated System Test (IST) is done, this will ensure that individual equipment tested will be able to operate as a one system. During the IST various Failure Scenarios are simulated as shown in Appendix A of this book.

Prior to IST it is normal to check Mechanical side of the system, which is known as Heat Load Test for which the Data Halls are fully loaded with design dummy loaders (normally 2-3kW hair dryers). The method statement for the Heat Load Test is shown in Appendix B of this book.

CHAPTER 2

Electrical System

Electrical Power and Lighting, Emergency Generators and Switchboards servicing the facility, with dual UPS source of power, consisting of the following components and sub-systems.

Photo 2.1 – Typical Dry type HV Transformer

Photo 2.2 Typical HV Switchgears

1) Transformers & HV Switchgears

   a) In some projects transformers and associated HV switchgear are provided by the utilities company. In that situation the utility company will carry out their own T&C works. But in some cases the client has to buy their own transformers and associated HV switchgears to speed up the construction period.
   b) The method statement for FAT and Site Acceptance test (SAT) will be very similar in contents which must be approved by T&C Consultant with the consent of MEP Consultant. All FAT and SAT should be carried out by High Voltage Registered Electrical Worker REW HV depending on local authorities.
   c) For T&C of transformers consultant will be responsible for full FAT and SAT for these transformers. Vendor will be required to submit their full method statements for both FAT and SAT. T&C of Transformers and HV Switchgears starts with physical checking of switchboards with approved drawings and checking of calibration certificates of all instruments to be used.

d) During the FAT/SAT of transformer the following tests are recommended: -

   i. General visual inspection of winding and tapping.
   ii. LV winding resistance measurement, acceptance criteria ≤0.3mΩ at 25 deg C
   iii. HV winding resistance measurement, acceptance criteria ≤2%
   iv. Insulation test HV winding to earth, acceptance criteria ≥5MΩ
   v. Insulation test HV winding to LV winding, acceptance criteria ≥5MΩ
   vi. Insulation test LV winding to earth. Acceptance criteria: ≥5MΩ
   vii. HV side Power Frequency Withstand Test 28kV for 60 seconds. Acceptance criteria: No damage to insulation. (FAT only)
   viii. LV side Power Frequency Withstand Test 3kV for 60 seconds. Acceptance criteria: No damage to insulation. (FAT only)
   ix. Induced AC withstands voltage test, applied voltage 760V, at 200 hertz for 30 seconds (FAT only).
   x. Measurement of no load loss current, to measure magnetising loss. Acceptance criteria compare with data sheet on name plate (FAT only)
   xi. Measurement of short circuit impedance and load loss. Acceptance criteria compare with data sheet on name plate (FAT only)
   xii. Partial Discharge measurement test, Pre-stress voltage of 19.8kV at 200Hertz for 30 seconds, acceptance criteria, ≤10pC (FAT only)
   xiii. Dry type transformers are equipped with ventilation fans and in this case a temperature sensor control panel will be fitted on the transformer enclosure which will indicate the core winding temperature of all three phases, Temperature

control panel will control the operation of ventilation fans. When the transformer reaches extremely high it will send signal to ACB and trip the circuit. Appropriate alarm signal will be sent to BMS.

e) During the FAT/SAT of HV Switchgears the following test are recommended: -

   i. Physical checking of switchboards with approved drawings.
   ii. Insulation test of switchboard before High Voltage Power Frequency (HVPF) test and after HVPF.
   iii. High Voltage Power Frequency (HVPF) Test,
   iv. Measurement of Voltage Ratio Test
   v. Primary Injection Test (Current Ratio Test),
   vi. Contact Resistance test acceptance criteria depends on the size of bus-bars
   vii. Secondary Injection Test
   viii. Testing of Under Voltage Relays

f) For T&C of HV cables the following tests are recommended: -

   i. Core cable identification test using continuity tester
   ii. Insulation resistance test before and after High Pressure Test, normally with 5kV or 10kV tester acceptance >100MΩ
   iii. DC high pressure test 25kV for 15 minutes' acceptance criteria of leakage current will depend on the length and type of cables

2) Emergency generators and associated fuel supply equipment

Photo 2.3 Stand by Generator Set

Photo 2.4: Typical Fuel System Pump Installation.

a) Gensets are installed to provide backup power during the utility company's power failure. It is to be noted during that the genset start up time the UPS will take up the load without break. In hotel projects many it may be just one generator covering both essential supplies and fire services. And small UPS units or built-in UPS will be installed to back up computer rooms
b) T&C of generator start with Factory Acceptance Test (FAT). This is normally attended by representatives from client, MEP Consultant, T&C Consultant, Main Contractor and genset Vendor. In many cases it will be an overseas trip unless the genset Vendor have full testing facilities locally. The main purpose of FAT is to ensure that the genset is fully in compliance with the specification prior to shipment. Normally the T&C Consultant will issue a FAT report to verify this immediately after the FAT to confirm the genset can be shipped to site. All FAT and SAT should be carried out by competent person from genset vendor.
c) The method statement for FAT and Site Acceptance test (SAT) will be very similar in contents which must be approved by T&C Consultant with the consent of MEP Consultant.
d) T&C of genset starts with physical checking of serial number, name plate with approved documents. Checking of calibration certificates of all testing equipment to be used.
e) The normally the following tests are recommended: -

   i. Insulation tests on all incoming and outgoing cables, acceptance criteria: $1M\Omega$ with 1000Vdc Megger
   ii. Control Function Test to test whether genset could be started and stopped in both manual & automatic modes.
   iii. Auto start and emergency stop test
   iv. Protection Test such as High engine temperature test criteria: 101 deg C ±3 deg C warning, 104 deg C ± 3 deg C engine shut down some manufacturer have lower temperature settings

v. Engine over speed test criteria: 1650rpm shut down for 50Herts.
vi. Low oil pressure test criteria: Warning 2.6 Bar ± 5%, Shutdown 2.2 Bar ± 5%. This depends on the size of engine
vii. Under-voltage test criteria, 342V shut down for 380V supply system for Hong Kong power supply declared voltage of 380V
viii. Over-voltage 418V shut down for 380V supply system. These are typical for a 2Mw Cummins genset rated at 380V, 3Phase & Neutral, 50Herts. It is to be noted that the different manufacturers will have different set values.
ix. Step load tests, 0-25%, 25-50%, 50-75%, 75-100%, 100-0%
x. After that the genset should ran for 7 hours continued at 100% followed by an hour at 110% load. During this time, it is important to note how the coolant temperature and the oil pressure vary with load. It is also required to note down the fuel consumption record so fuel consumption rate per hour can be calculated
xi. Another aspect on T&C of genset is to carry out a cold start on 60% load and measure the voltage dip and note the time taken to recover back to normal voltage. Acceptance criteria is normally ≤20% of declared voltage.
xii. Coolant low level shut down normally by simulation.
xiii. 3 Attempts fail to start test or cranking test. After 3 attempts the control panel should receive fail to start alarm.
xiv. Vibration measurement test, Acceptance criteria as per manufacturer's recommendation.
xv. Noise measurement test. Acceptance criteria as per manufacturer's recommendation.

f) For fuel system the following tests are recommend: -

   i. Welding test on at 20% of joints by Specialist Welding Testing firm. This test will be based on BS EN 1290: 1998 Magnetic Particle Examination of weld. Surface temperature of pipe works shall not exceed 50 Deg C and the acceptance criteria shall be in accordance with BS2971: 1991 section 7.
   ii. Hydraulic pressure test for all pipe works between the under-ground fuel tank and the genset. Fuel pipes are normally pressure tested at 700kPa for 2 hours.
   iii. Hydraulic pressure test on under-ground fuel tanks at 70kPa for 2 hours.
   iv. Insulation test for Fuel pump's electrical installation is required with acceptance criteria minimum of 1MΩ with 500Vdc megger.
   v. Fuel pump function test will include normal function test, like auto start, auto stop,
   vi. Auto change over between duty pump and standby pump
   vii. Lamp and buzzer test,
   viii. Emergency stop and overload test complete with interface tests with BMS.
   ix. Fuel tank low level alarm test
   x. Fuel tank high level alarm test
   xi. Above tests are complete with interface tests with BMS.

3) **Main incoming LV switch boards, MCC Switch boards, Genset Switch boards, UPS Switch boards**

Photo 2.5 Typical Low Voltage Switchboard

a) Low Voltage switchboards (LVSB) are fed direct from transformer output. From these LVSBs the electrical energy is distributed to Data Centres or other mostly installed data centres, or other floors in Hotel.

b) The method statement for LVSBs Factory Acceptance Test (FAT) and Site Acceptance test (SAT) will be very similar in contents which must be approved by T&C Consultant with the consent of MEP Consultant.

c) T&C of LVSB start with Factory Acceptance Test (FAT). This is normally attended by representatives from client, MEP Consultant, T&C Consultant, Main Contractor and UPS Vendor. In many cases it will be an overseas trip. The main purpose of FAT is to ensure that the UPS units are fully in compliance with the specification prior to shipment. Normally the T&C Consultant will issue a FAT report to verify this immediately after the FAT to confirm the UPS units can be shipped to site. All FAT and SAT should be carried out appropriate Registered Electrical Worker (REW). This will be either REW B or C depending the largest size of Air Circuit Breaker (ACB) for Hong Kong

d) T&C of LV Switchgears starts with physical checking of switchboards with approved drawings. Checking of calibration certificates of all testing equipment to be used.

e) In general, the following test are required:

   i. Insulation test of switchboard first with all ACBs open. Acceptance criteria 1MΩ with 1000Vdc Megger.
   ii. Insulation test of switchboard first with all ACBs close. Acceptance criteria 1MΩ with 1000Vdc Megger.
   iii. High pressure injection test normally 2.5kV for 60 seconds, acceptance criteria a maximum spill current of 100mA.
   iv. Current Transformer direction check
   v. Primary Injection Test, this involves connecting a variable current transformer (CT) on primary side of (CT) then increasing slowly until the primary side full current of CT ratio is achieved on secondary side (FAT only).
   vi. Measurement of contact resistance on major bus-bar connection also known as ductor test which is measured in μΩ. Switchboard manufacturer will publish the acceptance criteria for their switchboard.
   vii. Secondary Injection Test, normally on old IDMT relay, for earth fault test plug setting is set to 1A and time setting to 1 and for over current the plug setting is set to 5A and the time setting to 1.
   viii. Torque test on all bus-bar connection, Switchboard manufacturer will publish the acceptance criteria for their switchboard bus-bars bolt sizes.
   ix. Under-voltage relay test, normally 90% of declared voltage, over-voltage test normally 110% of declared voltage.

x. Some client requires LVSB temperature rise test which can take one full day for one switchboard. Acceptance criteria is maximum temperature rise of 80 degrees C.

a) During Primary Injection Test is normally carried out in FAT and same is not required to repeat in SAT. Typical High pressure injection test result is shown below: -

Test Terminals	Test Voltage	Duration	Leakage Current mA	Criteria in mA
E to L1 + L2 + L3 + N	2.5kV	60 Seconds	3.93	<100mA
N to L1 + L2 + L3 + E	2.5kV	60 Seconds	2.60	<100mA
L1 to L2 + L3 + N + E	2.5kV	60 Seconds	2.78	<100mA
L2 to L1 + L3 + N + E	2.5kV	60 Seconds	2.29	<100mA
L3 to L1 + L2 + N + E	2.5kV	60 Seconds	2.34	<100mA

b) The Typical test results from Secondary Injection Test is shown on the table below: -

Phase	Relay Setting		Multiple of Plug Setting	Injected Current	Operating Time (Seconds)		
	Plug Setting	Time Setting		Secondary	Nominal	Acceptable Range	Actual
L1	5	1		10			3.73
L2	5	1		10			3.67
L3	5	1	2	10	3.90	3.17-4.63	3.61
Neutral	1	1		2			3.59
L1	5	1		25			1.86
L2	5	1	5	25			1.89
L3	5	1		25	1.85	1.58-1.98	1.82
Neutral	1	1		5			1.96
L1	5	1		50			1.34
L2	5	1		50			1.32
L3	5	1	10	50	1.30	1.20-1.40	1.32
Neutral	1	1		10			1.37

Automatic Transfer Switch (ATS)

a) Normally ATS fitted in LVSB are tested as per part of LVSB. However, ATS installed on other part of Installation i.e. stand-alone ATS is tested separately. This SAT is to be carried out by competent person from ATS vendor
b) For SAT of ATS the following checks/tests are recommended: -

   i. Ensure that both Normal Supply A and Essential Supply B are available and both supply healthy indicator lights are 'On'.
   ii. Select the Normal Supply A as preferred source.
   iii. Switch 'Off" source A, within 3 to 5 seconds the load should transferred to source B.
   iv. Now switch 'On' back source A, within 3 to 5 seconds the load should transferred back to source A.
   v. Select the Essential Supply B as preferred source.
   vi. Switch 'Off" source B, within 3 to 5 seconds the load should transferred to source A.
   vii. Now switch 'On' back source B, within 3 to 5 seconds the load should transferred back to source B.

4) **Main Distribution Boards (MDB)/MCB Board**

a) T&C of MDB starts with physical checking of MDB/MCB board with approved drawings. Checking of calibration certificates of all testing equipment to be used. T&C for up to 400A can be carried out by REW 'A'
b) For SAT of MDB/MCB boards the following tests are recommended: -

   i. Insulation test with 500Vdc megger on incoming and outgoing cables: Acceptance criteria 1MΩ
   ii. Polarity test on incoming and outgoing cables: acceptance criteria ohms meter reading near 0Ω

iii. Live test on incoming cables, these include measurement of line and phase voltages, phase sequence test and earth impedance test: Acceptance criteria close to declared voltage for voltage, phase rotation clockwise.
iv. Continuity test of CPC: Acceptance criteria ohms meter reading near 0Ω
v. Continuity of Ring final circuits: Acceptance criteria ohms meter reading near 0Ω
vi. Earth loop impedance test on other final circuits. Acceptance criteria as per BS7671 depending on the size of cable
vii. RCCB tripping test on commando socket.: Acceptance criteria≤ 0.4s
viii. For large cables torque is also to be carried.

5) **Uninterrupted Power Supply (UPS) Units**

a) UPS units are mostly installed data centres, but not so much in hotels. In many data centre project will use dual UPS sources namely A source and B source. This way if one source fails it will be fed from the other source.
b) T&C of UPS units start with Factory Acceptance Test (FAT). This is normally attended by representatives from client, MEP Consultant, T&C Consultant, Main Contractor and UPS Vendor. In many cases it will be an overseas trip. The main purpose of FAT is to ensure that the UPS units are fully in compliance with the specification prior to shipment. Normally the T&C Consultant will issue a FAT report to verify this immediately after the FAT to confirm the UPS units can be shipped to site.
c) The method statement for UPS FAT and Site Acceptance test (SAT) will be very similar in contents which must be approved by T&C Consultant with the consent of MEP Consultant. FAT and SAT should be carried out by competent person from UPS vendor.
d) T&C of Units starts with physical checking of UPS unit with approved drawings. Checking of calibration certificates of all testing equipment to be used.

e) The standards regarding the UPS *(IEC 62040-3 and CEI ENV 50091-3)* provide a number of tests. The following tests that are considered by the standards but are not mandatory: -

   i. UPS Performance test
   ii. Load transfer test both automatic and manual (waveform to be captured)
   iii. AC main failure test and Return test (waveform to be captured)
   iv. Step load test 0-25%, 25-50%, 50-75%, 75-100%, 0-100%, 100-0% (waveform to be captured)
   v. Short circuit test (FAT only) UPS shut down within 5.5 seconds.
   vi. 125% overload test for 10 minutes, 150% overload test for 1 minute. (All alarms to be noted and actual time to be recorded with a stop watch). After the specified time the load transfers automatically to bypass mode.
   vii. Voltage Distortion with balanced load acceptance criteria: THD less than 1%, Phase displacement error less than 1%, Output voltage regulation less than 1%
   viii. Voltage Distortion with unbalanced load acceptance criteria: THD less than 1%, Phase displacement error less than 1.5%, Output voltage regulation less than 3%
   ix. Metering and Alarm test
   x. UPS Burn-in test for 8 hours at 100% load
   xi. A/C to A/C efficiency test at 25%, 50%, 75% & 100% load. At 100% load it should be not less than 90% efficiency
   xii. Output voltage Regulation not more than 1 at 100% load
   xiii. Record Harmonics not more than 5% for current, 1% for voltage at 100% load
   xiv. Record Power factor not less than 0.85 at 100% load
   xv. Load bus synchronised test (SAT only)
   xvi. Battery discharge test (Actual backup time in minutes depend on design) (SAT only)

f) The standards regarding the DRUPS *(IEC 62040-3 and CEI ENV 50091-3)* provide a number of tests. The following tests that are considered by the standards but are not mandatory: -

1. Typical DRUPS outdoor installation consists of three containers, namely DRUPS container largest unit which contains main engine complete with clutch, batteries, alternator, accumulator, daily fuel tank, fuel cooler, radiator fans, Exhaust fans. SAT involves: -

   a. Engine protection alarm tests for example high coolant temperature shut down, high fuel pressure shut down, low coolant level shut down, emergency shut down etc.
   b. Interface test with fuel system
   c. Accumulator recovery test
   d. NER test at 100% Load
   e. All control cable test point to point to COP
   f. Standalone 4-hour burn-in test at 100% load
   g. Unit vibration test
   h. Unit noise level measurement test
   i. Unit step load test at differing load conditions
   j. Unit unbalance load test L1=10%, L2-10% and L3-0% load
   k. Unit load sharing test in after connected to IP Ring
   l. Control logic test of radiator fans and ventilation fans

2. MV container which contains all MV switchgears, SCADA panel, all control cables from engine to MV panel and vice versa, all MV cables from main HV board to MV switchgears, all MV cables from MV switchgears to Choke container.

   a. SAT of MV switchgear similar to any other MV Switchgears
   b. SAT of all HV and LV cables
   c. MV switchgear interface test with SCADA and COP.
   d. SAT of Earthing system
   e. SAT of local DB board
   f. SAT of lighting and power system

g. SAT of F.S Installation
h. SAT of Split type air conditioning system

3. Choke container which contains both MV and IP chokes, all MV cables from MV containers, step up transformer, step down transformer for auxiliary power for three containers, COP panel which contains all electronic controls for the engine and PLC for MV switchgears and Profibus, Auxiliary panel and Neutral Earth Resistor.

   a. SAT of HV Cables from MV switchgear to Auxiliary Transformer, NB Choke and IP Choke
   b. SAT of LV Cables from DRUPS to Step-up Transformer
   c. SAT of Auxiliary Transformer - HV side high pressure test at 28kV for 1 minute, megger test at 10kV before and after the High pressure test. LV side high pressure test at 3kV for 1 minute, megger test at 1kV before and after the HP test.
   d. SAT of Step-up Transformer - HV side high pressure test at 28kV for 1 minute, megger test at 10kV before and after the High pressure test. LV side high pressure test at 3kV for 1 minute, megger test at 1kV before and after the HP test.
   e. SAT of No Break Choke - Insulation test with 10,000 Megger before and after the HP test. HP test at 28kV for 1 minute.
   f. SAT of IP Choke - Insulation test with 10,000 Megger before and after the HP test acceptance criteria >100MOhms. HP test at 28kV for 1-minute acceptance criteria <100mA.
   g. SAT of Neutral Earth Resistor Breaker
   h. Interface test with SCADA and COP
   i. SAT of Earthing system.
   j. SAT of local DB boards and final circuit
   k. SAT of Lighting and Power system
   l. SAT of F.S. Installation
   m. SAT of split type air conditioning system

Photo 2.6: A partial picture of DRUPS showing Accumulator and the Alternator

Photo 2.7 front view of DRUPS NB-Choke Unit

Photo 2.8: An elevation view of DRUPS MV Switchgears

6) Static Transfer Switch (STS) units

Photo 2.9 Typical STS unit black colour next to PDU grey colour

a) STS units are mostly installed data centres, but not so much in hotels. Inputs to STS dual sources namely preferred source and alternate source. This way if the preferred fails it will be fed from the alternate source.
b) T&C of STS units start with Factory Acceptance Test (FAT). This is normally attended by representatives from client, MEP Consultant, T&C Consultant, Main Contractor and STS Vendor. In many cases it will be an overseas trip. The main purpose of FAT is to ensure that the STS units are fully in compliance with the specification prior to shipment. Normally the T&C Consultant will issue a FAT report to verify this immediately after the FAT to confirm the STS units can be shipped to site.
c) The method statement for STS FAT and Site Acceptance test (SAT) will be very similar in contents which must be approved by T&C Consultant with the consent of MEP Consultant. These tests should be carried out by competent person from STS vendor.
d) T&C of Units starts with physical checking of STS unit with approved drawings. Checking of calibration certificates of all testing equipment to be used.
e) The standards regarding the UPS *(IEC 62040-3 and CEI ENV 50091-3)* provide a number of tests. The following tests that are considered by the standards but are not mandatory: -

   i. Input to preferred source 1 failure, load transfer to S2, acceptance criteria ≤5ms.
   ii. Input to preferred source 1 resumed, load transfer back to S1, acceptance criteria ≤5ms.
   iii. Input to preferred source 2 failure, load transfer to S1, acceptance criteria ≤5ms.
   iv. Input to preferred source 2 resumed, load transfer back to S2, acceptance criteria ≤5ms.
   v. Manual transfer load, acceptance criteria ≤5ms.

vi. Maintenance bypass load transfer test, acceptance criteria ≤5ms.
vii. Unsynchronised load transfer test, acceptance criteria ≤5ms.
viii. STS alarm test include, input failure, CB tripped, Current overload, under voltage, over-voltage, SCR short circuit (FAT only), SCR open circuit (FAT only)
ix. A/C to A/C efficiency test, normally over 99% efficiency

7) Power Distribution Units (PDU)

Photo 2.10 Front view of typical custom made PDU unit

a) PDU units are mostly installed data centres, but not so much in hotels. Inputs to PDU are normally single source and marked either A or B source
b) T&C of PD units start with Factory Acceptance Test (FAT). This is normally attended by representatives from client, MEP Consultant, T&C Consultant, Main Contractor and PDU Vendor. In many cases it will be an overseas trip. The main purpose of FAT is to ensure that the PD units are fully in

compliance with the specification prior to shipment. Normally the T&C Consultant will issue a FAT report to verify this immediately after the FAT to confirm the PD units can be shipped to site.

c) The method statement for PDU FAT and Site Acceptance test (SAT) will be very similar in contents which must be approved by T&C Consultant with the consent of MEP Consultant. These tests should be carried out by competent person from PDU vendor.

d) T&C of Units starts with physical checking of PD unit with approved drawings. Checking of calibration certificates of all testing equipment to be used.

e) The standards regarding the PDU (ISO 9001, ANSI, IEEE, NEPA-70 and NEPA-75) provide a number of tests. The following tests that are considered by the standards but are not mandatory: -

   i. Insulation resistance test, acceptance criteria $1M\Omega$
   ii. High pressure test (FAT only), acceptance criteria 30mA
   iii. CT direction test (FAT only)
   iv. CT ratio test (FAT only), acceptance criteria ± 5%
   v. Power meter accuracy test, acceptance criteria ±1.5%
   vi. Contact resistance test
   vii. Inrush current check, acceptance criteria 6xFLC
   viii. MCCB function test including over load test
   ix. Insulation test on incoming cable
   x. Insulation test on final circuits.
   xi. Live test on final circuits including RCD/MCB

8) **Main Cables**

   a) T&C of main cables starts with physical checking of cable schedules on approved drawings. Checking of calibration certificates of all testing equipment to be used

b) For SAT of Main cables, the following tests are recommended: -

   i. Cable polarity check. Acceptance criteria ohm meter reading near 0.
   ii. Insulation test with 500Vdc megger. Acceptance criteria 1MΩ
   iii. Torque test after the cable termination. Acceptance criteria as per LVSB manufacturer's recommendation.
   iv. Phase rotation check. Acceptance criteria clockwise
   v. Measurement of phase voltages and line voltages. Acceptance criteria of ±6% of declared voltage.
   vi. Measurement of earth loop impedance. Acceptance criteria as BS7671

9) **Lighting System**

   a) Lighting in Data centres are installed as per (Telecommunication Industry Association) TIA-942-A guide, lighting fittings are installed above the isles and between the IT cabinets. Lighting are provided in 3 levels:

      i. Level 1: Data centre in unoccupied mode
      ii. Level 2: Initial Entry to Data Centre
      iii. Level 3: Data centre in fully occupied mode, illumination requirement normally 500 lux in horizontal plane and 200 lux in vertical plane.

   b) The lighting in Data centres will normally be controlled by occupancy sensors and monitored by BMS. The main purpose behind this is for energy saving. When a motion is detected by the motion sensor, the lighting zone controlled by particular sensor will be switched on. When this motion is inactive for 15 minutes depending on the time delay setting on the sensor, the lighting will be switched off.

c) Lighting system in hotel will be slightly different, where functional lighting is required for external architectural lighting, Lobby lighting, Restaurant lighting, Ballroom lighting, Bar lighting, Meeting rooms, Guest room lighting and Spa lighting.
d) Lighting control for Guest rooms are via through a local MCB boards. Lighting will be locally controlled via many switches to provide different lighting scene to suit guest's comfort.
e) For SAT the following tests/checks are recommended: -

   i. Insulation test on final circuit cables with 500Vdc megger. Acceptance criteria 1MΩ
   ii. Final circuit polarity test on final circuit. Acceptance criteria near zero on Ohm meter reading
   iii. Earth loop impedance test on final circuits. Acceptance criteria as per BS7671
   iv. Lux level measurement with all normal lighting switched on. Acceptance criteria as per TIA-942-A
   v. Lux level measurement with all emergency lighting on battery and all normal lighting switched off for two hours. Acceptance criteria 2 lux
   vi. Function test of occupancy sensors. Acceptance criteria as per TIE-942-A
   vii. Interface test with BMS

10) Earthing System

a) In Data Centres there will be two distinct earthing system namely electrical and clean earthing. Former for normal low voltage system and latter for IT system. Some MEP Consultant prefers a separate earth for genset system as per BS7671. In addition to these there will be separate earthing network for the building lightning protection system.
b) T&C of earthing should start at very start of building construction when locations of main earth pits and lightning

earth pits are being built. It is easier to measure the resistance of earth electrodes when external part of building is not concreted. As there will be enough space to drive test spikes.

c) For actual T&C the following tests are recommended: -

i. Measurement of resistance for main earth electrodes.
ii. Measurement of resistance for lightning earth electrodes.
iii. Measurement of resistance for clean earth electrodes.
iv. Continuity test of earth tape between earth pit and transformer.
v. Continuity test of earth tapes between transformer and LVSB
vi. Continuity test of CPC between LVSB and MDB/MCB boards.
vii. Measurement of earth loop impedance test at LVSB incomer.
viii. Measurement of earth loop impedance test at MDB/MCB board incomer
ix. Measurement of earth loop impedance at final circuits

11) SAT of Lighting Motion Sensors

i) Ensure all motion sensors are installed as per approved drawing showing all zones
ii) Ensure they are all labelled
iii) Ensure all sensors are damage free
iv) Check there is adequate maintenance access is available
v) Set all motion sensor time delay on the relay say 10 seconds
vi) Move out other occupants from the room and close all access doors
vii) Stand still in one corner of the room and wait until all lights are off automatically.
viii) Start walking towards motion sensors from one end of the room until the first sensor on that zone is activated.

ix) Continue above until all sensors are successfully activated and the room light is fully lit.
x) Now stand still again until all room lights are automatically switched off.
xi) Now switch on/off the room lighting using the switch
xii) Record these on the test record sheet.

CHAPTER 3

Mechanical Ventilation and Air-conditioning System

The Air Conditioning and Mechanical Ventilation Systems servicing the facility consisting of the following components and sub-systems.

Photo 3.1 Typical Water Cooled York Chiller

1) **Water-cooled Chiller /Air cooled chiller**

   a) In both Data Centres and Hotels will have a number of chillers. actual numbers will depend on the size of project and the total cooling load required.

   b) After the award of contract, the Chiller vendor will submit method statements for both FAT and SAT. This should include chiller protection tests, Emergency stop, Chiller performance test at 100% full load, Noise measurements and chiller starts up reports. This method statement may be similar for FAT and SAT which must be approved by the T&C consultant with no adverse comments from the MEP consultant

c) T&C of Chillers starts with physical inspection of, checking of serial number and model numbers, checking of calibration certificates of instruments to be used for the test. The main purpose of FAT is to ensure that the Chiller units are fully in compliance with the specification prior to shipment. Normally the T&C Consultant will issue a FAT report to verify this immediately after the FAT, to confirm the Chiller units can be shipped to site.

d) For FAT/SAT the following tests are recommended: -

　　i. Motor insulation test, acceptance criteria 1Ω
　　ii. Carry out initial start and note cooler pressure drops and condenser pressure drops
　　iii. Ensure the oil pump is running and take note of pump current
　　iv. Check both default and set value for Base Demand limit, LCW set point, ECW set point, Ice built set point, tower fan set point if any.
　　v. Check and note down both default and set value on VFD Configuration sheet.
　　vi. Check and note down both default and set value on Equipment service (Lead/Lag) configuration sheet
　　vii. Check and note down both default and set value on Equipment Service(Ramp-Dem) configuration sheet
　　viii. Check and note down both default and set value on Equipment Service (Temperature Control) configuration sheet
　　ix. Measure the compressor inrush current and voltage drop during the start up.
　　x. Note down the compressor demand percentage and record Compressor current in all three phases.

xi. Take note of motor voltage, Oil pump voltage and control voltage
xii. Activate emergency stop and ensure it is working correctly
xiii. High discharge pressure refrigerant cut out test, acceptance criteria for 30XA1202 Air cooled chiller is 1900kPa.
xiv. Low suction refrigerant temperature protection cut out test, acceptance criteria for 30XA1202 Air cooled chiller is -28 degrees C.
xv. Low oil protection pressure protection test, acceptance criteria for 30XA1202 Air cooled chiller is 2kPa
xvi. Evaporator freeze protection test, acceptance criteria for 30XA1202 Air cooled chiller is -1.3 degrees C.
xvii. Emergency stop test,
xviii. Chiller performance test at 100% full load
xix. Noise measurement

2) **Cooling Towers (CT)**

Photo 3.2 Cooling Tower

a) Cooling Towers are installed for water cooled chillers. The size of cooling tower varies according the cooling capacity of chillier.
b) FAT for cooling tower is normally not required. Therefore, the vendor for cooling tower will submit SAT method statement which will be approved by T&C consultant after MEP consultant have no major comment.
c) T&C of CT starts with physical inspection of CT, checking of serial number and model numbers, checking of calibration certificates of instruments to be used for the test.
d) For SAT the following tests are recommended: -

   i. Insulation test on fan motor circuits.
   ii. Check the fan rotation, belt tension, drive alignment and fan blade installation. Turn the fan by hand and ensure free rotation. Adjust the belt and align the motor.
   iii. Check and inspect the drain, overflow and equalizing pipe connection. Adjust the water basin level and float movement. Remove any accumulated dirt or sediment to prevent any sediment from entering the system.
   iv. Check and inspect the drain, overflow and equalizing pipe connection. Adjust the water basin level and float movement. Remove any accumulated dirt or sediment to prevent any sediment from entering the system.
   v. Check and adjust all the valves to an appropriate position.
   vi. Start up the condensing pump only and make sure no water sprays out the tower.
   vii. Inspect the water distribution inside the cooling tower.
   viii. Check the bleed-off control by testing the conductivity of the water.
   ix. Inspect the drift eliminator performance.
   x. Turn the fan motor on. Check the motor electrical performance and motor rotation speed.
   xi. Check the function of the Variable Frequency Drive (VFD) and ensure that it is within the required Hertz.
   xii. Check the thermal performance of the cooling tower.

xiii. Check any unusual noise and vibration on full and half speed.
xiv. Check the fan motor overload function and verify setting
xv. Using an electronic manometer, measure the water flow rate at the DRV of the cooling tower.
xvi. Measure entering and leaving air temperatures (DB & WB) using a sling psychomotor
xvii. Measure entering and leaving water temperatures using a manometer
xviii. Record all data on test sheets

3) **Chilled water pumps (CWP)**

Photo 3.3 – Typical Chilled Water Pump Installation

a) Chilled waters between the chillers and the loads via the main chilled water risers, nowadays it is common that these pumps are control by Variable Frequency Drive (VFD) to save energy.
b) FAT for CWP is normally not carry out, CWP vendor will provide their FAT data during the SAT

c) Therefore, the vendor for CWP will submit SAT method statement which will be approved by T&C consultant after MEP consultant have no major comments.
d) T&C of CWP starts with physical inspection of CWP, checking of serial number and model numbers, checking of calibration certificates of instruments to be used for the test.
e) Ensure pump alignment test is completed as stated in following steps: -

   i. Place the dial gauge on the coupling or shaft on the driver side (Pump side)
   ii. Attach the tactile needle on the rim surface of the pump coupling or (motor coupling) and turn both couplings together gradually
   iii. Set the dial gauge to zero and turn both sides of the motor and pump coupling together for 90 degrees. Get the reading of the dial gauge for parallel offset
   iv. Turn the coupling for another 90 degrees. Get the reading of the dial gauge for parallel offset.
   v. Turn the coupling for another 90 degrees. Get the reading of the dial gauge for parallel offset.
   vi. Finally turn the coupling back to the 0-degree position.
   vii. Attach the tactile needle on the rim surface of the pump coupling or (motor coupling) and turn both couplings together gradually.
   viii. Set the dial gauge to zero and turn both sides of the motor and pump coupling together for 90 degrees. Get the reading of the dial gauge for angular offset
   ix. Turn the coupling for another 90 degrees. Get the reading of the dial gauge for angular offset.
   x. Turn the coupling for another 90 degrees. Get the reading of the dial gauge for angular offset
   xi. Finally turn the coupling back to the 0-degree position.
   xii. The tolerance of the parallel offset and angular offset shall be within the tolerance as stated in the specification (0.05 mm

xiii. If the parallel and angular offset is not within the stated tolerance, adjust the shims on the pump accordingly and carry out the alignment as stated above until the offset is within the tolerance

f) In water cooled chiller system, the secondary pump system should be isolated for this test. For SAT the following checks are recommended: -

   i. Measurement points are suitably positioned
   ii. System is thoroughly tested, watertight, thoroughly flushed & cleaned.
   iii. All strainers have been removed, cleaned and replaced.
   iv. System is filled, vented and water treatment has been applied
   v. System pressurization unit is operational.
   vi. Cold feed valve is open.
   vii. Pump bearings and all external parts are clean
   viii. Components are secure, impeller is free to rotate and flow direction is correct.
   ix. Pump/Motor couplings are secure and correctly aligned
   x. Motor and Pump are lubricated with the correct grade of lubricant.
   xi. Glands are packed & adjusted to correct drip rate
   xii. Motor and drive guards are fitted access is available for tachometer.
   xiii. Check that all normally open isolating and regulating valves are fully open and that all normally close valves are closed.
   xiv. Check that all control valves of branch circuits are opened
   xv. Check that all AHU etc. control valves are opened.
   xvi. Pump casing is vented of air
   xvii. Pump suction valve is open, standby pump isolation valves are closed and discharge/flow valve is 50% closed to limit initial start current.

xviii. Initial start and check
xix. Confirm that the rotation direction of the motor shaft is correct.
xx. Confirm that the motor, pump and drive are free from vibration and undue noise.
xxi. Record the motor starting current
xxii. Confirm the motor running current on all phases to ensure that they are balanced between phases.
xxiii. Confirm that there is no overheating of the motor
xxiv. Gradually open the discharge valve until the motor current reaches either the design value or the motor full load current, whichever is the lower.
xxv. Check the pump pressure developed by means of the pump altitude gauges against the design pressure. If excessive pressure is developed at this stage, the cause should be investigated.
xxvi. Adjust the discharge valve so that the flow as determined from the pump characteristic is 100 +10/-0% (per cent) of the design value. Note that the motor full load current shall be not exceeded.
xxvii. Pump performance test and checks
xxviii. Water flow rate and supply pressure (suction and discharge). Flow should be within -0% to +10% of the design valves. The flow rate can be measured by using an Electronic flow meter at main metering station or orifice plate.
xxix. Note Motor/Pump rotational speed.
xxx. With Pump Closed Head (note Zero Flow suction & discharge pressures).
xxxi. With Full open head (note Operational Flow suction and discharge pressures).
xxxii. Record Motor Running Current of each phase and voltage.
xxxiii. Record all these data on the test sheets

4) **Condensate water pumps**

   a) SAT of condensate pump is same as that of chilled water pumps described above
   b) Pump Alignment Test

      1. Place the dial gauge stand on the coupling or shaft on the driver side.
      2. Attach the tactile needle on the rim surface of the pump coupling or (motor coupling) and turn both couplings together gradually.
      3. Set the dial gauge to zero and turn both sides of the motor and pump coupling together for 90 degrees. Get the reading of the dial gauge for parallel offset
      4. Turn the coupling for 90 degrees. Get the reading on the dial gauge for parallel offset.
      5. Turn the coupling for 90 degrees. Get the reading on the dial gauge for parallel offset
      6. Finally turn the coupling back to the 0-degree position
      7. Attach the tactile needle on the rim surface of the pump coupling or (motor coupling)
      8. Set the dial to gauge to zero and turn both sides of the motor and pump coupling together for 90 degrees. Get the reading of dial gauge for angular offset.
      9. Turn the coupling for another 90 degrees. Get the reading on the dial gauge angular offset
      10. Turn the coupling for another 90 degrees. Get the reading on the dial gauge for angular offset
      11. Finally turn the coupling back to the 0-degree position
      12. The tolerance of parallel offset and angular offset shall be within the tolerance of 0.05mm as stated in specification
      13. If the parallel and angular offset is not within the stated tolerance, adjust the shims on the pump accordingly and carry out alignment as stated above until the offset is within the tolerance

5) **Chemical dozing pumps**

   a) Chemical dozing pumps are installed near the chiller

Photo 3.4 A General view of chemical dosing pumps

   b) Chemical dozing pumps are installed to for treatment of condensing water and chilled water.
   c) These pumps are designed to inject certain amount of chemical for fixed duration depending on the quality of water.
   d) Initial checking includes visual checking of P.E. tanks, pipe works
   e) Function check of Electricity supply, ensure the Circuit breaker is working correctly
   f) SAT of these includes function test of duty and standby pumps.
   g) Function check of power supply indicator
   h) Metering Pump No. 1 and 2
   i) Low level alarms on chemical container.
   j) Operation check of PLC controller for scheduled operation

6) **Separator pumps for cooling tower Cleaning System**

   a) Carry out Pump Alignment test as described in Condensate Pump above
   b) The main purpose of these pumps are to ensure that the basin of cooling towers is kept clean at all time by flushing out dirt at regular intervals.
   c) In LAKOS system about 32 nozzles are in tower basin to promote cross flows on the basin of cooling towers.
   d) SAT of Separator pump will include measuring the current and voltage of motor when running
   e) Operation check of time schedule, operation time can be adjusted

7) **CRAC units**

   a) Visual Check

      i) Ensure that the Unit is clean
      ii) Ensure that all bolts and fixing are secure
      iii) Ensure that the condenser fan is free to rotate
      iv) Ensure the fan belt is in normal condition without wear and tear
      v) Check the pulley alignment
      vi) Check the belt tension
      vii) Ensure the air filter is clean
      viii) Ensure that the condensate drain flows smoothly
      ix) Make sure there is no leaking in coil, piping, valves and fittings
      x) Ensure the heater is fixed and secure
      xi) Ensure the thermostat is working correctly.
      xii) Ensure that the chilled water pressure gauge and temperature sensors are working properly

xiii) Ensure that water supply for humidifier is ready
xiv) Ensure that all indications on the control panel are normal
xv) Ensure the buzzer is working properly
xvi) Take down the detail of Filter sizes

b) Start-up and Testing Procedure

i) Switch on the CRAC unit
ii) Record the incoming voltage (370 – 400V for Hong Kong)
iii) Record the control power voltage (23 – 25V)
iv) Press the Start button on the key pad to start the CRAC
v) Ensure the rotation of fan is correct
vi) Measure the ambient temperature and humidity
vii) Record the running current of motor No.1 & 2
viii) Switch on the heater No.1 and measure the current
ix) Switch on the heater No.2 and measure the current
x) Set and turn the RH valve to lowest setting
xi) Measure the current of humidifier
xii) Measure the total air flow on the return side
xiii) Ensure the RH and the temperature is as per design value
xiv) Check filter clog alarm

c) Operation Measurement Electrical

i) Measure room temperature
ii) Measure RH
iii) Check Supply voltage
iv) Check control voltage
v) Heater Current
vi) Humidifier Current
vii) Chilled water pressure Supply and Return
viii) Chilled water temperature Supply and Return
ix) Ensure cooling valve is working properly
x) Take down the detail on regulating valve

d) Air Flow Measurement

   i) Measure the air flow on 4 points in return side
   ii) By calculation work out the total air flow rate and compare with the design value.
   iii) Clearly write down the % of design value

e) Interface Test with other CRAC Units

   i) Switch on all CRAC units
   ii) Trip the duty CRAC unit
   iii) The standby unit should switch on automatically
   iv) Initiate Fire Trip all CRAC units should trip
   v) Finally, with all CRAC units running, there is a requirement to measure air flow in each floor grilles, this is known as floor grille balancing

Photo 3.5 – Typical CRAC Installation

8) **SAT of Fan Coil Units**

   a) Pre-operation check

      i) Ensure that the FCU is installed properly
      ii) Ensure that the thermostats are correctly installed
      iii) Ensure that the modulating valve is working and no leakages are found
      iv) Open the washable filter and ensure all are clean
      v) Check that motors are lubricated and securely bolted
      vi) Examine the fan impeller make sure it is clean and free to rotate
      vii) Make sure fresh air supply to FCU is not obstructed.
      viii) Megger the Fan motors and ensure that the reading more than 1 MΩ

   b) Function Check

      i) Switch on the FCU and observe for any abnormal noise
      ii) Ensure there is no vibration
      iii) Check that the water Drain pan can be drained out smoothly
      iv) Measure the voltage and operating current
      v) Measure the total air flow and compare with the design value and record it.
      vi) Measure the both supply air and return air temperature
      vii) Finally press E-Stop to check it is functioning properly

9) **SAT of Exhaust Air Fans**

   a) Insulation Test

      i) Carry out Megger test using 500V Megger
      ii) The reading should be more than 1M Ohms

b) Operation Result

   i) Measure the motor voltage
   ii) Measure the motor starting current
   iii) Measure the motor running current
   iv) Write down the type of starter i.e. DOL
   v) Check the overload setting of the motor circuit
   vi) Measure the rpm of the motor if possible
   vii) Measure the flow rate and compare with the design value

c) Local Motor Control Panel (LMCP) Function Test

   i) Start the fan manually then stop
   ii) Start the fan and E-Stop it
   iii) Start the fan and activate overload device
   iv) Start the fan and Fire trip if applicable
   v) Check the buzzer function and mute

d) Interface test with BMS

   i) Activate common alarm and ensure it can be received in BMS
   ii) Check the fan status

10) SAT of Fresh Air Fans

a) Sometimes Fresh Air Fans are installed in some areas. SAT for this kind of fans are exactly same as Exhaust Fans described in 9) above.

11) SAT of Chillers Control System (CCS)

Ser#	Control Requirement Description	Test Procedure	Acceptance Criteria
A	**Chiller Control – Normal Start up**		
A.1	Check that all chiller plant equipment in auto mode	Selector Switched to auto mode on the LMCP	Received Auto in BMS system
A.2	Check Chiller/ Equipment normal operation status	Check chiller set generated any alarm from equipment or LMCP	Selected next chiller/ equipment
A.3	Start up the chiller program	Send command to start chiller plant program	Receives the command from BMS workstation
A.4	CHW and CDW Cooling Tower valve open	Automatic command and start the relative valve control	a. Valves opens and receive the feedback correctly b. After time delay and feedback if the feedback signal is not received, changeover to next equipment
A.5	Start CHWP, CDWP, and Cooling Tower fan	Automatic command start the relative equipment control	a. All the relative equipment starts and receive the running status. b. After time delay and feedback if the feedback signal is not received, changeover to next equipment
A.6	Start the Chiller Plant	Automatically command to start the relative equipment control	All the relative equipment start and receive the running status

B.	**Chiller Stage up/ Stage down and Weekly Rotation**		
B.1	Stage up by Building Load. Simulate the load above the set point	Change the set point for Chiller plant program>default setting	After the time delay of 300 seconds, Receive the command (demand+1) from BMS workstation
B.2	The least running hour chiller set will command to start	Command the program start for chiller plant program	Next Chiller set start up
B.3	Stage down by Building Load. Simulate the load below the set point	Change the set point for chiller plant program<default setting	After the time delay of 300 seconds, Receive the command (demand-1) from BMS workstation
B.4	The longest running hour chiller set will command to stop	Command the program stop for chiller plant program	Next chiller will stop
B.5	Stage up by supply temperature. Simulate the temperature above the set point	Change the temperature set point from 9 Deg C to 6 Deg C	After the time delay of 300 seconds, Receive the command (demand+1) from BMS workstation
B.6	The least running hour chiller set will command start	Command the program start for chiller plant program	Next Chiller set will stop
B.7	Stage down by supply temperature. Simulate the temperature below the set point	Change the temperature set point from 6 Deg C to 9 Deg C	After the time delay of 300 seconds, Receive the command (demand-1) from BMS workstation
B.8	The longest running hour chiller set will command to stop	Command the program stop for chiller plant program	Next chiller will stop
B.9	Stage up by chiller running current. Simulate the current above the set point	Change the set point for Chiller plant program>default setting	After the time delay of 300 seconds, Receive the command (demand+1) from BMS workstation
B.10	The least running hour chiller set will command start	Command the program start for chiller plant program	Next Chiller set will start

B.11	Stage down by chiller running current. Simulate the current above the set point	Change the set point for Chiller plant program>default setting	After the time delay of 300 seconds, Receive the command (demand_1) from BMS workstation
B.12	The longest running hour chiller set will command to stop	Command the program stop for chiller plant program	Next chiller will stop
B.13	Trigger the chiller weekly rotation function, Stage up the chiller	Command the program start for chiller plant program	Next Chiller set will start
B.14	Stage down after time delay	Command the program stop for chiller plant program	Next chiller will stop
C.	**Chiller Set Fault changeover Control**		
C.1	Chilled water pump trip by activating the running pump trip alarm	Trip Alarm generated from the LMCP	Receives trip alarm
C.2	Stop the chiller set and start up next chiller plant	Command the program stop for chiller plant program	Next chiller will start and current one stop
C.3	Condensing water pump trip by activating the running pump trip alarm	Pump trip alarm generated on the LMCP	Receives trip alarm
C.4	Stop the chiller set and start up next chiller plant	Command the program stop for chiller plant program	Next chiller will start and current one stop
C.5	Chilled water pump trip by activating the running pump trip alarm	Trip Alarm generated from the LMCP	Receives trip alarm
C.6	Stop the chiller set and start up next chiller plant	Command the program stop for chiller plant program	Next chiller will start and current one stop
C.7	Cooling tower trip by activating the running pump trip alarm	Trip Alarm generated from the LMCP	Receives trip alarm

C.8	Stop the chiller set and start up next chiller plant	Command the program stop for chiller plant program	Next chiller will start and current one stop
C.9	Valve trip by activating the current pump trip alarm	Trip Alarm generated from the LMCP	Receives trip alarm
C.10	Stop the chiller set and start up next chiller plant	Command the program stop for chiller plant program	Next chiller will start and current one stop
D.	**Chilled water pump speed control**		
D.1	Decrease pump speed control	Close all CRAC, PAU and AHU common Valve	The chilled water pump speed decrease to achieve the pressure set point
D.2	Increase the pump speed control	Open all CRAC, PAU and AHU chilled water Valve	The chilled water pump speed increase to achieve the pressure set point
E	**Condensing water pump speed control**		
E.1	Decrease pump speed control	Adjust the set point for the flow rate from 80 l/s to 60 l/s	The condensing water pump speed decrease to achieve the flow rate of 60 l/s
E.2	Increase the pump speed control	Adjust the set point for the flow rate from 60 l/s to 80 l/s	The condensing water pump speed increase to achieve the flow rate of 80 l/s
F.	**Free Cooling Pump/ Heat Exchanger Control**		
F.1	Check that free cooling pump is in normal operation	Check if LMCP has generated any alarm	Pump alarm generated from LMCP and exit the free cooling mode
F.2	Decrease pump speed control	Adjust the set point for the flow rate from 80 l/s to 60 l/s	The condensing water pump speed decrease to achieve the flow rate of 60 l/s
F.3	Increase the pump speed control	Adjust the set point for the flow rate from 60 l/s to 80 l/s	The condensing water pump speed increase to achieve the flow rate of 80 l/s

F.4	PI control open heat exchanger valve	Adjust the chilled water return temperature set point from 8 Deg C to 10 Deg C	The valve will modulating open to change the return temperature achieve the water temperature set point of 10 Deg C
F.5	PI control close heat exchanger valve	Adjust the chilled water return temperature set point from 10 Deg C to 8 Deg C	The valve will modulating close to change the return temperature achieve the water temperature set point of 8 Deg C
G.	**Bypass Valve/ Bleed off valve**		
G.1	PI control open heat exchanger valve	Manually close the riser valve to increase the water pressure	The modulating valve will be modulating open to change pressure achieve pressure set point
G.2	PI control close heat exchanger valve	Manually open the riser valve to decrease the water pressure	The modulating valve will be modulating close to change pressure achieve pressure set point
G.3	Control open the bleed off valve	Adjust the set point to <800 ppm	The bleed off valve will open
G.4	Control close the bleed off valve	Adjust the set point to >600 ppm	The bleed off valve will close
H	**Cooling Tower Speed Control**		
H.1	Increase the cooling tower Fan Speed control	Adjust the set point from 40 Deg C to 35 Deg C for the condensing water	The fan speed will increase to change the condensing water temperature to achieve the set point
H.2	Increase the cooling tower Fan Speed control	Adjust the set point from 35 Deg C to 40 Deg C for the condensing water	The fan speed will decrease to change the condensing water temperature to achieve the set point

H.3	Stop the cooling tower fan	Change the fan speed at the minimum speed of 25Hz	After the time delay of 180 secs the fan will be off
H.4	Adjust the set point for the condensing water temperature	Command the program parameter for the chiller plant	Bypass valve will modulating open
I.	**Free cooling/partial free Cooling mode**		
I.1	Change the chiller plant into partial free cooling mode or free cooling mode.	Adjust the set point 1(default 9 Deg C)> the ambient wet bulb and the set point 2(default 12 Deg C)>condenser water temperature	After the time delay of 180 secs, chiller plant change to partial free cooling mode or free cooling mode.
I.2	Change the chiller plant into partial free cooling mode	Adjust the set point 3(default 10 Deg C)> the ambient wet bulb and the set point 4(default 16 Deg C)>condenser water temperature	Chiller plant activate to partial free cooling mode and change the set point 4 to condenser pump set point.
I.3	Free cooling mode activate	Adjust the set point 3(default 10 Deg C)> the ambient wet bulb and the set point 4(default 16 Deg C)>condenser water temperature	Chiller plant activate to partial free cooling mode and stops the chilled water pump.
J	**Chiller Plant Demand Limit Control**		
J.1	Reduce the chiller demand limit	Adjust the building load set point<the average of chiller load	After the time delay of 180 secs, reduce the chiller demand limit.
J.2	Release the chiller demand limit	Adjust the building load set point<the average of chiller load	Release the chiller demand limit to 100%
K	**Chiller supply temperature point reset**		
K.1	Adjust the CWS temperature set point	Check all cooling valve position <80-100%	After the time delay of 300 secs, increase the CWS temperature set point 0.5 Deg C.
K.2	Release the CWS temperature set point	Check all cooling valve position >80-100%	Release the CWS temperature to default

12) SAT of Air Handling Units (AHU)

Photo 3.6 Typical AHU Installation

a) Starting up Fan Motor, Switch on the machine
b) SAT of AHU include measurement of motor pulleys to ensure it complies with the specification.
c) Measurement of motor starting and running current
d) Measurement of motor speed
e) Function test of LMCP panel
f) Measurement of air flow in all air grilles and checking against the designed value
g) Measurement of air temperatures at air grilles in air conditioned space
h) Physical inspection of bag filters
i) Test Fire Trip operation to ensure that AHU is trip off automatically if a smoke detector is activated in the air conditioned space.
j) Check that filter clock alarm is functioning normally
k) Check the temperature of both CHWS and CHWR

l) Check that VRV valve setting and water flow to the AHU is as per design value.

m) Measure the chilled water flow rate on Supply and ensure it matches with the design flow rate

13) SAT of Primary Air Handling Units (PAU)

a) The SAT of PAU is exactly same as that of AHU as described above with the exception that suction duct of PAU is connected to Fresh Air Grille outside.

b) In Hotels it is necessary to ensure that PAU is working properly as in some cases this is the only source of fresh air supplies to Guest Rooms which is very important.

14) SAT of Split Type Air-conditioning Units

a) Outdoor Unit

  i) Ensure that all pipe work is completed
  ii) Carry out pressure test of 440 PSI for the duration of 24 hours using Nitrogen gas
  iii) During the operation mode press E-Stop to ensure it is working

b) Indoor Unit

  i) Switch on the unit and measure the voltage and current
  ii) Measure the total air flow from the indoor unit check with design value
  iii) Set the local thermostat to 16 Degrees C
  iv) Let it cool the room for few minutes then take the Room temperature
  v) Take the supply air temperature at air supply grille
  vi) Take the return air temperature at return air grille
  vii) Ensure that the expansion valve opening is normal

15) SAT of Water Leak Detection System

a) Installation

   i) Ensure that the installation is completed power is available.
   ii) Ensure that the water leakage detection map is displayed next to panel

b) Function test

   i) Simulate leakage on 1 meter and ensure the alarm is activated
   ii) Check the leakage distance shown on the panel is correct
   iii) Disconnect the first cable connector and cable break alarm is activated leakage on 35 metres and ensure the alarm is activated.
   iv) Check the leakage distance shown on the panel is also about 35m
   v) Ensure this is repeated in BMS panel
   vi) Repeat i) to vi) above for remaining installation

16) Chilled Water Balancing

a) Pre-check

   i) Ensure that the system is fully flushed and cleaned
   ii) Verify the installation of the valves as per design
   iii) Ensure the pump head has sufficient capacity to meet the static pressure for new installed chilled water system
   iv) Ensure the regulating devices and the other components such as double regulating valve, gate valves etc. are within the pipe works
   v) Ensure that all NO/NC valves are in correct status
   vi) Visual checking of water tightness of the whole system

b) Test Procedure

   i) Manually open all Independent Balancing Valves of CRAC units and FCU. The system bypass line should be fully closed
   ii) The system is assumed to be balancing since an independent balancing valve is provided for all CRAC, FCU and PAU. The required water flow rate should be pre-set by and achieved by these independent valves. Conceptually, double regulating valves should be measured flow rate only and should be fully opened.
   iii) Measure the water flow rate by water flow meter at the DRV at the branch off for Riser 1 and Riser 2 of each floor.
   iv) Chilled water balancing will be completed with new equipment on Chilled Water Riser 1, and then repeat with all new equipment on Chilled Water Riser 2
   v) Adjust the zone flow valves to design flow rate of plus or minus 10% for all zone valves.
   vi) Take down the record of this measurement on test record sheet
   vii) Measure the water flow rate at DRV of all CRAC, FCU and PAU on whole floor.
   viii) Take down the record of this measurement on test record sheet
   ix) Reset all Balancing valve in auto mode.
   x) The acceptance tolerance level of chilled water flow rate to be plus or minus 10% for all zone valves.
   xi) The main chilled water pipe branches DRV will be set according to the load design.

17) Duct Leakage Test

a) Preparation

   i) Identify and agree section of duct work to be tested and requirement for blanking plates
   ii) Ensure there are no obvious leaks in the duct work or any damage
   iii) Identify the ductwork classification and required test pressures

b) Test Execution

   i) Blank off any open ends in ductwork test section
   ii) Calculate the surface area of ductwork test section and identify maximum permitted leakage from the specification
   iii) Connect test fan to ductwork test section and set up manometer
   iv) Start the fan and adjust the fan flow rate until the required static pressure differential is achieved
   v) Check the measured leakages is within the permitted range
   vi) Maintain the test for a minimum of 15 minutes and after this period check the measured leakage is not increase
   vii) Switch off the fan and allow the static pressure in the duct reduce to zero and then restart the fan and check that the leakage has not increased

18) Variable Air Volume) VAV System

a) VAV system is used in modern high rise offices. All VAV boxes are connected on the Supply Air Duct from the floor PAU normally by a flexible duct.
b) The outlet air from these VAV boxes are distributed to the air conditioned space by boot diffusers. Then returns back to PAU via Return Air Duct

c) The amount of air flow into the space is directly controlled by Damper Position which in turn is controlled by the feedback room temperature. Higher the room temperature more opening up to Vmax and lower the room temperature less opening up to V min.

d) It is important to measure the total Air Flow from the PAU. This is done by measuring the air flow rate at 4-6 pivot points depending on the size of the duct then finally calculating the total air flow using the total cross section area of the Supply Air duct. It is to be noted during this time all VAV boxes must be set to V max by BMS.

e) SAT of VAV Boxes

   i) Ensure that the flow measuring hood and pitot are available.
   ii) Ensure that electricity is already installed by electrical contractor
   iii) Verify that all VAV zoning is completed and programmed accordingly.
   iv) Switch on the AHU/PAU
   v) Set all the VAV box to be tested to V max i.e. 100%
   vi) Measure the air flow from each VAV boot outlets and take down the record on the test record sheets
   vii) Continue above for all VAV boxes until all VAV boxes on the floor served by related PAU are completed
   viii) Now add up the total of all VAV boxes, this should match with the total air flow measured in d) above
   ix) Now set all VAV boxes in V min
   x) Repeat above ii) to iv) and compare the results
   xi) It is normal to control the room temperature set point by installing occupancy sensors.
   xii) When the occupancy sensor is off, the VAV set point is set to 28 degrees C, so the VAV box will set automatically to V min thus less energy is wasted.

xiii) On the other hand, when the occupancy sensor is on, the VAV set point is set to 24 degrees C, so the VAV box will open more in order to cool down the room to achieve the set point of 24 degrees C.

19) Flushing and Cleaning of Chilled Water Pipes

a) Preparation Works

i) All chilled water pipework installations shall be complete and have been successfully pressure tested prior to the start of the flushing and cleaning process.

ii) A specialist flushing and cleaning subcontractor or parties familiar with flushing & chemical dozing should carry out the flushing and cleaning.

iii) Permanent water connections have been made or temporary water supply is sufficient for the purpose of flushing.

iv) All air vents, dosing pots and large bore flushing valves necessary for proper venting, dosing and draining of the system shall be installed and operational. Calculating pumps size (Or temporary pumps where necessary) and pressurization units/make up tanks shall be operational.

v) For pipework 50mm bore or below, specific flushing drain valve that are to be utilized during the flushing process shall be line size. For pipework above 50mm bore, these valves shall be 65mm. They shall be installed across all major plant items, at the base of all risers, at the end of headers and any other location that will facilitate flushing. Should it be necessary to install other, typically larger drain valves these shall be installed on a temporary basis and removed once the flushing and cleaning process has been successfully completed.

vi) Temporary or permanent flushing loops shall be provided across coils, 2way valves and any component that might become blocked or damaged during the flushing and chemical cleaning process. Such bypassed or blanked items shall be flushed individually with dosed water on the reconnection with the system.

vii) If section of the pipework system are required to be pre-operationally cleaned separately, all necessary bypass connections and loops shall be provided to ensure positive circulation through the section to be cleaned

viii) Prior to the start of flushing, the cleaning specialist shall inspect the installation and high light any potential problem areas such as possible low velocity sites with no drain points that compromise the flushing process. The contractor shall undertake any remedial works that are necessary to ensure that the flushing system can be flushed both forwards and backwards

ix) All flow restricting devices (flow Self Regulating Valves) must have their flow regulating cartridges removed from the system prior to commencing flushing

b) Flushing Procedure

i) All strainers shall be removed, cleaned and replaced during the flushing and cleaning process and immediately before water balancing. Advance notice of agreed duration, shall be given to the client's team prior to removal and cleaning of any strainers,

ii) The contractor or chemical cleaning specialist shall ensure that the water velocity in the system is sufficient to provide adequate collection of all debris in horizontal pipework in accordance with table below 3.1 The contractor and cleaning specialists must ensure that the minimum recommended velocities are achieved and maintain for each system or circuit. Measurement of water flow should be taken using a calibrated manometer or ultrasonic flow meter.

Nominal Size of Pipe (mm)	Flushing Velocity (m/s)	Flushing Volume (l/s)
15	0.96	0.20
20	1.00	0.37
25	1.03	0.60
32	1.06	1.08
40	1.08	1.49
50	1.11	2.45
65	1.15	4.25
80	1.17	5.98
90	1.19	8.10
100	1.21	10.47
125	1.24	16.41
150	1.26	23.98

Table 3.1,: Minimum Flushing Velocity and Volume Requirement

iii) During the flushing and cleaning process the specialist contractor shall maintain a log events detailing progress on each stage showing time scale and current quality of water such as pH, TDS and Fe.

iv) All stages of flushing will be witnessed by T&C consultant

v) Flushing and cleaning process will continue until the T&C consultant is satisfy that the system has been cleared of all debris and contaminating matter.

vi) On completion of flushing and cleaning process the system shall be filled with clean water and chemically dosed to inhibit corrosion and bacterial / algae growth. A final set of water sample is to be taken immediately prior to practical completion to ensure the satisfactory quality of water.

vii) No hydronic system shall be left empty once the flushing and cleaning processes have been completed

viii) On completion of the cleaning process all flush/drain valves shall be capped or plugged. Bypasses shall be valve off or removed as required

20) Air Balancing of Floor Mounted Diffusers

a. Pre-commissioning Checks

   i)   Check the design volumes serving each floor plenum
   ii) Floor void inspected for integrity around the edges ensuring there are no paths for air to scape other than floor diffusers.
   iii) All floor tiles are in place within the pressurized zone
   iv) All CRAC units within the prepared plenum zone are ready to run
   v)   All supply air diffusers are fully open. Remove grille to ensure damper that damper basket is also fully open.
   vi) Ensure that all individual CRAC units have been fully commissioned and operational.
   vii) CRAC unit air return paths & unit enclosure are clear and unobstructed.
   viii) All air filters are clean and in position
   ix) Record the CRAC name plate information and compare with specification and highlight any discrepancy noted.

b. Floor Grille Proportional Balance

   i)   With all CRAC units within plenum zone operating and correct supply air temperature carry out the following.
   ii) Draw a schematic layout indicating the location of floor grilles within plenum zone together with CRAC orientation
   iii) Scan all floor grille outlets and fill out detail in fully open column of test sheet
   iv) Balance floor outlets using a calibrated measure hood. The maximum tolerance between all grilles to be within 20% of index grille
   v)   With all grilles suitably balanced finally scan the grilles and write down these records on the test form.

vi) Finally scan return air inlet grilles to CRAC units using an anemometer to determine the total system air volume. Divide this air volume by the number of grilles to determine the correction factor for the measuring air hood. Record these on the test sheet.
vii) Once the floor grilles have been balanced, the team shall move onto the next operational condition, i.e. with different units operating and different units on standby. Each floor grille will then be scanned again to ensure that the air flow rate is still to the design.
viii) Each operation scenario will be tested in accordance with vii) above with different units in a run condition to compare the results.

CHAPTER 4

Fire Services System

The Fire Services Systems servicing the facility consisting of the following components and sub-systems.

a) AFA System

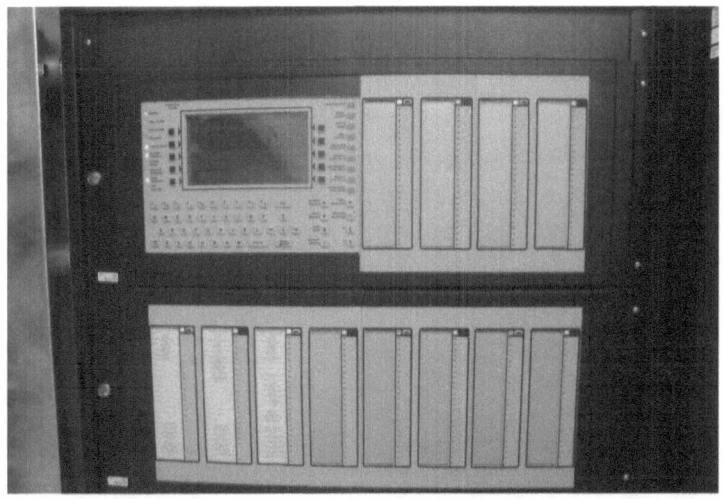

Photo 4.1 Typical AFA Panel

SAT of AFA system

    i) Ensure that all detectors are installed as per approved drawings
    ii) Activate each detector as per drawing and ensure it receives alarm in AFA panel and the red Fire lamp is lit
    iii) All floor internal bells are sounded
    iv) Ensure appropriate zone is indicated
    v) Appropriate text messages the device address and location is shown

vi) Ensure all flashing lights are flashing
vii) Appropriate BMS signal is received
viii) Like pre-action alarm
ix) Pre-action Water release
x) Pre-action alarm fault
xi) FM200 alarm
xii) FM200 Gas Discharge
xiii) FM200 Panel Fault
xiv) Remove detector head and ensure detector fault alarm is received
xv) Replace the detector head and ensure the fault alarm is cleared
xvi) Disconnect the AC power to AFA panel, AC fault alarm will activate
xvii) Disconnect battery power to AFA panel, battery fault alarm will activate in BMS panel

b) **FM200 System**

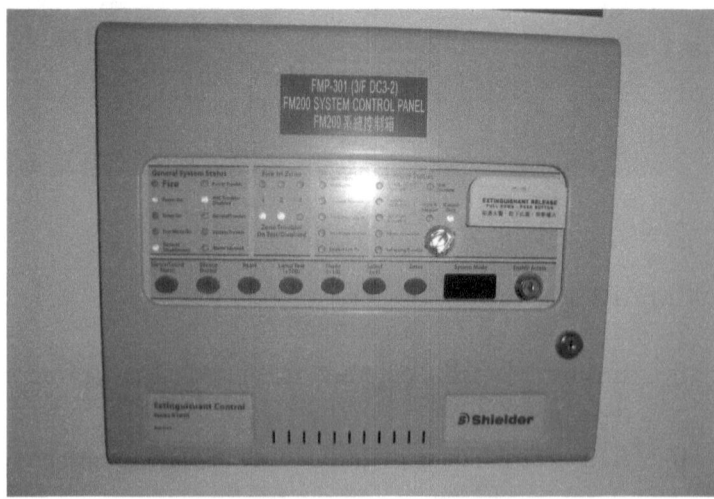

Photo 4.2 FM200 Control panel

This system is normally used in Data Hall. Main SAT of this is described as follows: -

Pressure Test and Puff Test

    i)   Pressure test for Gas pipes normally 10 minutes on 150 PSI
    ii)   Gas Medium N2 without any pressure drop
    iii) Write down the size of Gas Pipe and number Nozzles
    iv) Connect the pipe to N2 gas and open the valve
    v)   Observe continuous gas flow on the nozzle then turn off

Function Test in Auto Mode

    i)   Ensure that all personnel are notified about the test
    ii)   Ensure the system is isolated and actuator is removed
    iii) On the Gas Control Panel select Auto mode
    iv) Activate first detector in Zone 1(Stage 1)
    v)   Alarm will be sent to Building AFA Panel
    vi) Observe audible and visual signal are activated
    vii) All local bells are sounding.
    viii) All Flashing lights are flashing.
    ix) Activate second detector in Zone (Stage 2)
    x)   30 Seconds count down start
    xi) Observe more audible and visual signals are activated
    xii) Flashing lights are flashing
    xiii) All sirens are sounding
    xiv) After 30 seconds solenoid will activate
    xv) Gas will be released in the space.
    xvi) VAC will trip activate
    xvii) On the control panel reset the fire Alarm

Test in Manual mode

    i)   On the Gas Control Panel select Manual mode
    ii)   Activate first detector in Zone 1

iii) Alarm will be sent to Building AFA panel
iv) Observe audible and visual signals are activated
v) Bells are sounding
vi) Flashing lights are flashing
vii) Activate second detector in Zone 2
viii) The extinguishing system will not operate.
ix) Pull down the manual release switch
x) Local Bells & Siren are sounding
xi) Gas Discharged Do Not Enter sign operate
xii) On the Control Panel Reset the Alarm.

Fault Check Test

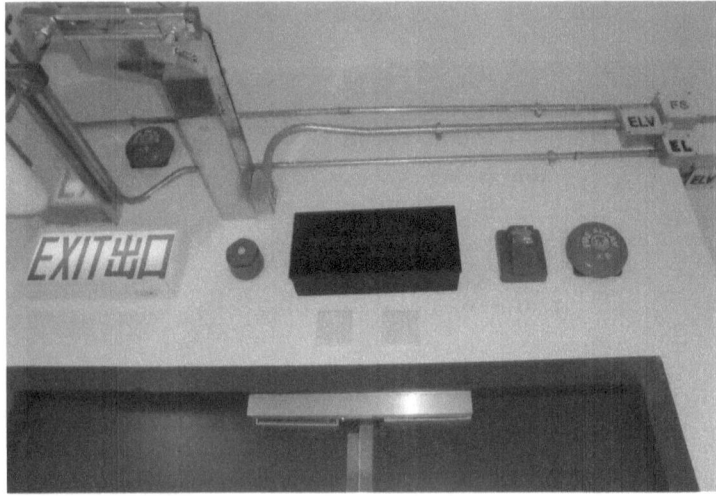

Photo 4.3 FM200 Alarm bells, Flashing Light, Siren and Notice

i) Remove the detector head from Zone 1
ii) Detector fault in Zone 1 will indicate on the control panel
iii) Remove the detector head from Zone 2
iv) Detector fault in Zone 2 will indicate on the control panel
v) Switch off AC supply, power supply failure alarm appears

c) Sprinkler System

SAT of Sprinkler System

i) Carry out Pressure test on the whole system 2 hours 15 bars
ii) Acceptance criteria no reduce in pressure
iii) Close subsidiary valve to check micro switch operation at panel.
iv) Check correct address of micro switch matches at fire panel
v) Re-open the subsidiary valve and confirm micro switch reset.
vi) Confirm test drain is sized to simulate a single sprinkler head activation.
vii) Connect hose to test valve and drain in a suitable location.
viii) Open test drain valve to simulate sprinkler head activation.
ix) Confirm flow switch activate and address of location is correct at fire panel
x) Confirm pump starts automatically
xi) The pressure and flow rate for high hazard on duty point and 140% duty point will be recorded.
xii) The pressures and its High/Low nominal flow for OH3 shall be recorded
xiii) Setting the pressure relief valve on DN65 tank return pipe shall be tested and recorded
xiv) Check that water motor alarm gong is audible
xv) Residual pressure at the most unfavorable location shall be recorded which is applicable to OH3/High Hazzard/In-Rack.
xvi) Close the test drain and confirm pump is still running
xvii) Manually re-set the pump – confirm sprinkle pump has stopped.
xviii) For each system tested, open a test valve within the plant room gradually to drop pressure. Note pressure Jockey pump cut-in, record this on the test sheet.

xix) Close the drain valve within the plant room and the Jockey pump stop. Note down the jockey pump cuts-out pressure on the record sheet
xx) Check and record system pressure from the installed gauges.
xxi) With the duty pump running, switch off the pump on the LMCP by tripping the motor overload switch. The stand by pump shall start. Note down the change over time on the test sheet.
xxii) With the duty pump running, switch off the pump by pressing the emergency stop switch. The standby pump should start. Note down the change over time on the test sheet

d) Hose Reel System

SAT of Hose reel system

i) Activate the break glass unit next to the hose reel
ii) Fire is send back to the building AFA panel
iii) Fixed pump will operate
iv) Alarm bells and Flashing light operate
v) Open the valve to let water spray come out
vi) Ensure that it can throw water zet up to 6m
vii) Reset the alarm and resume normal

e) Hydrant System

SAT of Fire Hydrant System

i) Pressure test the whole system for 2 hours at 185 PSI
ii) Acceptance Criteria no reduction in pressure.
iii) Carry out the SAT of Fixed pump as per other pumps
iv) Activate the fixed pump system
v) Connect the hose reel and open the water

vi) Measure the pressure at the end of hose reel, acceptance criteria 3.5 – 8.6 bars
vii) The water flow should be at least 450l/Min for a single hydrant.

f) VESDA System

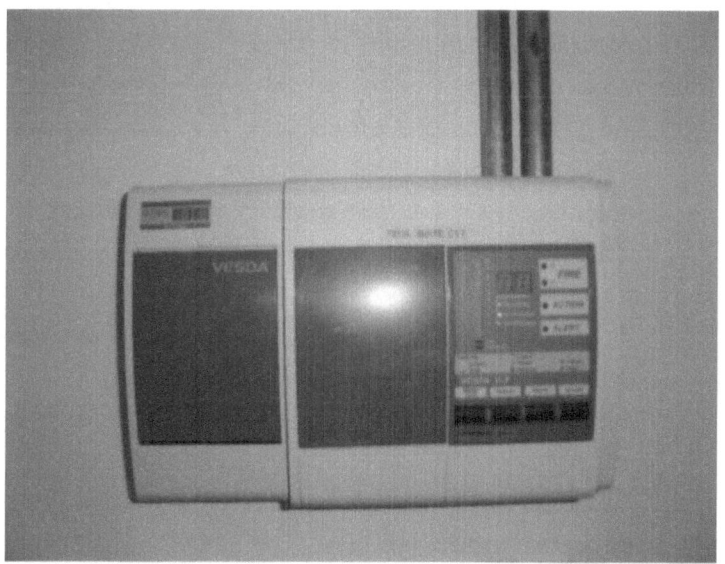

Photo 4.5 VESDA control panel

Visual inspection of VESDA System

i) Ensure that sampling pipe-works are installed as per approved drawing.
ii) Ensure the equipment installed matched with approved model
iii) Ensure that the layout of pipe nozzles is tally with approved drawing.
iv) Ensure that the smoke sensor level is properly adjusted

SAT of VESDA system using hot-wire test method

i) The test wire samples should exactly measure and cut to the required length
ii) Only use the recommended test wire. Under no circumstances should test wires be 1 m in length, of 10/0.1mm strands insulated with PVC to a radial thickness of 0.3mm
iii) It will be helpful to first coil the wire using a pen or pencil as temporary and then lay it out
iv) Alternately an arrangement to support the wire may be used with the precaution of a fire blanket positioned beneath the wire
v) Simulate fire condition at the remotest end cap of each zone on each system and record the response time of alert signal within 120 seconds.
vi) If the equipment is in a high flow area, then it may be required to shield the wire from the airflow.
vii) High current is passed through a test wire in order to heat it up, which causes its insulation to burn and gives off smoke. A timer is incorporated to provide a selectable burn period of up to 2 minutes
viii) The rocker switch on the top of the unit is the on/off switch. When the power is applied and the unit is turn on, the green Power indicator on the timer flashes and the power is applied to the output terminals until the pre-set timer has elapsed. Once the timer has timed out, the output power is switched off and the red "out" indicator on the timer illuminates. The power indicator then stops flashing and remains illuminated. The timer is reset when the switch is turned off and will run for the full selected period when the switch is turned on again.
ix) The unit must be operated for not more than 2 minutes continuously and must be left off immediately after operating for the period of 7 minutes to allow cool.

x) Within these 2 minutes all Alert stage, stage 1, stage 2 and Fire alarms are activated. Ensure this alarm is received in BMS

Testing Procedure using Gas (Aerosol Smoke)

i) Simulate a fire condition by test gas and randomly checking the sampling points at each zone on each system and record the response time of Alert signal within 120 seconds
ii) Ensure this alarm is received in BMS system

Alarm Check

i) Block the sample airflow inlet, filter clog alarm should be activated and send to BMS system
ii) Disconnect power supply and Battery cables, Power failure alarm should be activated.

g) Pre-action Double Lock Sprinkler system

Photo 4.6 Typical Pre - action Sprinkler Valves control system

SAT Procedure

i. Slowly open the main system drain valve. Note the pressure gauge, the air pressure drops to about 5-6 PSI, the compressor cut-in, and start automatically.
ii. Close the main system drain valve, the air pressure rises and compressor stop automatically
iii. Now open the drain valve slowly
iv. Compressor starts
v. Further open the system drain valve
vi. Note the supervisory signal
vii. Further open the main drain valve
viii. Zone 2 low air pressure alarm appears
ix. Close the system drain valve
x. Note that Air compressor will stop
xi. Close main water supply control valve, placing the circuit out of service (performing this results in operation of the deluge valve. Failure to close the main water supply control valve will cause water to flow into the sprinkler piping)
xii. Activate a smoke detector from zone 1 there should be no alarm on the pre-action control panel
xiii. Activate the second smoke detector in in zone 2 alarm will be activated in pre-action control panel. This will cause air pressure drop, the valve will open and the water will be discharged into dry system pipes.
xiv. To test the mechanical water motor alarm, open the test valve in the flow control water trim. The local water gong should be audible.
xv. After the above test close the alarm test valve
xvi. Ensure that supply piping to water alarm gong is properly drains
xvii. Verify that the alarm shut off valve is open, and the alarm test valve is closed.

xviii. Ensure that the outlet chamber is free of water. No water should flow from the drip check when the plunger is pushed

xix. Now record the pressure reading from the water supply pressure gauge.

xx. Verify that outlet chamber of the control valve is free of water. No water should flow from the drip check valve when the plunger is pushed.

xxi. Fully open the test valve

xxii. When a flow is detected from the flow test valve record the residual pressure from the water supply pressure gauge

xxiii. When the test is completed, slowly close the flow test valve

xxiv. Now simulate loss of battery power, check the pre-action valve spring should be close. Also repeat this test with loss of main AC power

xxv. To test pre-action valve release, this will depend on the activation of both electrical and pneumatic release. For actuation of electrical first release, first the deluge valve does not open unless the air pressure drops to a range of about 25 PSI in the system

xxvi. It is to be noted that pneumatic release is manual and mechanical and electrical release is automatic

xxvii. For activation of pneumatic release first the deluge valve will be opened.

xxviii. Pre-action valve does not release when two cross zone smoke detectors activate until the pressure drops and a low pressure alarm is received

xxix. Ensure that all air dampers on all protected zones are closed.

xxx. For actuation of manual release, deluge valve would open and causes water discharge into the protected areas.

CHAPTER 5

Plumbing and Drainage System

The plumbing and drainage systems servicing the facility consisting of the following components and sub-systems shall be tested as follows:-

1) **T&C Procedure for Cold Water Installation**

    a) Pre-requisite for SAT of Plumbing installation

    i) All pipework installation is completed
    ii) 2% welding test on all pipe works are completed
    iii) On pipe works greater than 300mm diameter pipe 5% welding test is completed
    iv) Pressure test all pipe works for 24 hours at 150 PSI
    v) Ensure the system is fully tested and no reduction in pressure

    b) Pre-Commissioning Check of Water Distribution System

    i) Divide pipe-work section into self-draining section
    ii) Isolate bypass items which are sensitive to dirt such pumps and bore coils.
    iii) When makeup or feed tanks are used for flushing ensure that the max possible pressure is sustained on the system during the flushing process
    iv) Flushing should continue until the outflow runs clear
    v) All pressure tests have been carried out and passed
    vi) Pump alignment test is completed
    vii) All strainers are physically cleaned and put back into the pipe-work
    viii) All pump impellers are free to rotate

ix) Anti-vibration mountings are checked for correct deflection
x) Alignment of all pulleys and couplings are correct
xi) The belt tension is correct as per specification
xii) The coolant is available at the bearing when specified
xiii) That the drive guards are fitted and the access for the tachometer is provided.
xiv) Internal wiring in pump control panels are properly installed
xv) Point to point test with for LMCP with BMS is completed
xvi) Power cables for the pumps are tested and securely installed

c) SAT of up-feed pump

i) Measure the electrical Voltage on the incoming side of the cable and ensure it is within the declared voltage of that country.
ii) Start the duty pump in manual mode and measure both starting and running current
iii) Measure the speed of the motor and compare with the name plate.
iv) Check the overload current setting
v) Repeat i) to iv) for Standby pump
vi) Set the Selector switch to Auto mode and start the duty pump.
vii) Now activate the overload by pressing the overload button in LMCP
viii) Duty pump should stop and the standby pump will start automatically
ix) Now reset the overload alarm on duty pump and press same on standby pump, the standby pump will stop and duty will start
x) While the duty pump is running activate the low level alarm, the pump will stop in low level cut-out
xi) Activate the high level alarm in the tank, same will be repeated in LMCP

xii) Reset the high level alarm in the tank, same alarm in LMCP will clear

xiii) Press lamp test button, all lamps should lit up

2) Drainage Installation

a) Water Pressure on Foul Water U/G Drainage Installation SAT Procedure

   i) Pressure test of 1.5m head is applied at high end of the pipe-work under test, while the test pressure at low end shall not exceed 6m head. Steeply grated pipes shall be tested by dividing into section. Test duration will be 30 minutes.

   ii) Remove all obstruction, debris and other matters from the pipe works

   iii) Secure all drain stoppers or bags on the end of pipe works and associated valves under the test

   iv) Fill the water to the pipe-works at least 2 hours before the test to allow water absorptions by the system

   v) At the beginning of test record the test pressure at the high end and the low end of the pipe works

   vi) After 30 minutes record the pressure at high end and low end again to calculate the loss.

b) Smoke Leakage Test (Applicable to pipe more than 300mm in diameter)

   i) Seal up the pipe work at both ends with blind caps

   ii) Inject the smoke into the pipe work using a smoke generator

   iii) No Leakage is allowed

c) Air Pressure Test (Applicable to pipe less than 300mm in diameter)

   i) An air pressure of 100mm shall be applied for the duration of 5 minutes

ii) Remove all obstruction, debris and other matters from the pipe works
iii) Seal the end of pipe lines and associated branches under test by drain plugs, inflatable canvas or rubber test bags.
iv) Connect U-Tube manometer to the pipe works.
v) Inject an air pressure of 100mm of water to the pipe works 5 minutes prior to test to allow stabilization of the air temperature and pressure inside the pipe-works.
vi) Take record of this air pressure at the beginning of test
vii) After 5 minutes take record the air pressure again to calculate the loss pressure.
viii) The acceptance criteria is a loss of no more than 25mm of water pressure within 5 minutes

d) SAT Procedure, Foul Water Drainage Installation above ground system

i) Seal the lower end of pipe works to be tested with plugs
ii) Connect a manometer to the pipe works
iii) Fill the pipe works with water to flood level of the lowest sanitary appliance at least 5 minutes before the test to see if any leakages of water is observe
iv) The static head shall not exceed 1.2m at the high point of the test and a maximum of 2.4m at the low point
v) The acceptance criteria is, it should not have any leakages of water below the lowest sanitary appliance

e) Air Pressure Test for pipe work less than 300mm in diameter)

i) During this test pipe works will be filled with air pressure of 38mm of water for a duration of 5 minutes for each drainage stack at the level above the lowest sanitary appliance
ii) All sanitary appliances to be fully charged with water seal

iii) Seal the ends of all pipe with plugs to all pipe works under test
iv) Connect U-tube manometer to the pipe works
v) Inject an air pressure of 38mm of water to the pipe works at least 5 minutes before the test to allow for stabilization of the air temperature and pressure inside the pipe
vi) Now take record of air pressure to start the test
vii) After 5 minutes take the air pressure record again
viii) Calculate the pressure loss, acceptance criteria is 10mm of water in 5 minutes.

f) Functional Performance Test

   i) During this test water will be discharged from the selected drain points to demonstrate the actual operation condition that water can be discharged simultaneously into the foul water drainage system during the normal operating condition. Visual check on the whole drainage system will be carried out for any back flowing. This will verify the design of foul drainage system
   ii) Remove all obstruction and debris from all drain point and pipe works
   iii) Discharge the water into the selected drain points from the water storage vessels simultaneously.
   iv) Carry out visual checks on the whole drainage system for any back flowing

g) SAT of Surface Water Drainage Installation U/G system, Pressure Test

   i) During this test pipe works are filled with water for the duration of 30minutes. Then pressure losses are calculated
   ii) A test pressure of 1.5m head shall be applied at high end of the pipe-work under test, while the pressure at the low end

shall not exceed 6m head. Steeply grated pipe work shall be tested by dividing into section.

iii) Remove all debris and obstruction from the pipe work.

iv) Install all drain stoppers and bags in the end of pipe work and associated branches under test.

v) Fill the water to the pipe works at least two hours before the test to allow for water absorption into the system.

vi) Now take record of pressure at high end and low end

vii) After 30 minutes take the record of pressure at high end and low end and calculate the pressure loss

h) Smoke Tight Test (for pipe work greater than 300mm in diameter)

i) During this test pipe works are sealed at both ends

ii) Smoke is injected into the pipe-works by external smoke generator.

iii) Acceptance criteria all pipe works must be smoke tight

i) Air Pressure Test (for pipe work greater than 300mm in diameter)

i) During this test all pipe works will be filled with Air Test Pressure of 100mm of water for 5 minutes.

ii) Remove all debris and obstruction from the pipe works under test

iii) Seal the ends of all pipe works and associated branches under the test by drain plugs, inflatable canvas and rubber test bags

iv) Connect U-tube manometer to the pipe work.

v) Inject an air pressure of 100mm of water to the pipe work at least 5 minutes before the test for stabilization of the air temperature and pressure inside the pipe works.

vi) Now take the record of air pressure inside the pipe works.

vii) After 5 minutes take the record of air pressure inside the pipe works again and calculate the loss

viii) Acceptance criteria will be that within 5 minutes the head of water should not fall by more than 25mm of water.

j) Leakage Test for Manhole

  i) During this test the manhole will be filled with water. The fall in water inside the manhole will be recorded and the compare with the maximum permissible fall. It should be noted that the manhole should be filled with water at least 20 hours prior to test for absorption period.
  ii) Remove all debris and obstruction from the manhole
  iii) Seal the ends of all pipe connection of the manhole by drain plugs, inflatable canvas or rubber test bags
  iv) Fill the water to the manhole at least 20 hours prior to test for absorption period
  v) Now measure the water level to the manhole to start the test
  vi) After 30 minutes re-measure the manhole water level inside the manhole and compare the results.
  vii) Acceptance criteria is no leakage

k) SAT of Surface Water Drainage Installation above ground system

  i) During this test each drainage stack will be charged with water at the level below the lowest sanitary appliances to see if any leakages of water can be observed.

l) Water Test of Surface Drainage Installation above G/L

  i) Seal the lower end of the pipe work being tested with plug
  ii) Connect a manometer to the pipe work
  iii) Fill the pipe work with water to flood the level of lowest sanitary appliance at least 5 minutes before the test to see if any leakage of water is observe.
  iv) Static head shall not exceed 1.2m at the high point of test and a maximum of 2.4m at the low end

m) Air Pressure Test of Surface Drainage Installation above G/L

    i) During this test pipe works will be filled with an air pressure of 38mm of water for a duration of 5 minutes

    ii) First fully charge with water seals on all sanitary appliances to be tested

    iii) Seal the ends of all pipe works with plugs at the pipe works being tested

    iv) Connect u-tube manometer to the pipe work

    v) Inject air to the pipe works at least 5 minutes before the test to allow stabilization of the air temperature and pressure inside the pipe.

    vi) Now measure the air pressure inside the pipe works and take down the record.

    vii) After 5 minutes measure the air pressure again and compare the results. Acceptance criteria is not more than 10mm of water pressure drop.

n) Functional Performance Test of Surface Drainage Installation above G/L

    i) During this test, water will be discharged from the selected drain points to demonstrate the actual operation condition that the water will be discharged simultaneously into the foul water drain system in actual operation condition. Visual check of whole system will be carried out for any back flowing. The result will be used to reflect whether the capacity of the foul water drainage system is adequate.

    ii) Remove all obstruction and debris from the drain points and pipe works

    iii) Discharge the water into the selected drain point from the water storage vessels simultaneously.

    iv) Now carry out visual check on whole drainage system for any back flowing

o) SAT of Gravity Condensate Drains

   i) Ensure that all pipe works are installed with gradient
   ii) Pour a bucket of water at upper end and the same is received at lower level. After this point either it is connected to the building main drain pipe or in some cases pump out using the sump pump as shown in 3) below.

p) SAT of Floor Drains

   i) Ensure that all floor drains are as per approved drawings
   ii) Pour a bucket of water in each drain point and ensure the water flows down smoothly.
   iii) Continue above until all floor drains are completed.

3) Sump Pump Installation

Photo 5.1 Typical Sump Pump Installation.

a) SAT of Sump Pump

   i) Take down the sump pump details
   ii) Carry out Insulation test on the cables
   iii) Ensure all level sensors are installed correctly
   iv) Test Pump 1 & 2 in manual mode
   v) Measure running current of both pumps
   vi) Test pumps 1 & 2 in Auto mode
   vii) Start pump No.1 and stop it pump No. 2 should start up
   viii) Stop pump No.2 No.1 should start up
   ix) Set up No. 1 in duty and No.2 in standby mode
   x) While No.1 pump is running add more water to create high level alarm No.2 pump should kick in.
   xi) Start pump No.1 and press overload device, Pump 2 should start
   xii) Start pump No. 2 and press overload device, Pump 1 should start.
   xiii) Carry out lamp test

CHAPTER 6

Building Management System

a) The Building Management System servicing the facility consisting of the following components and sub-systems.
b) DDC panels, Building Network Controllers, High Level Interface Units, patch panels, Edge switches, Core switches and printers
c) The Power Management System servicing the facility consisting of the following components and sub-systems. (Sometimes it is possible to have a separate monitor for the chiller control system)

    i) PMS server and workstation
    ii) Digital Multifunction Power Meters (CVM) and Digital Power Analyzers
    iii) Typical Data Centre will have BMS Interface Test with the following sub systems. All of these interface tests should be completed prior to Integrated System Test

a) Transformers: -

    i) High Temperature Trip

b) Generators: -

Ser.	Alarm	Received in BMS	Comments
1	Common Fault		
2	Gesnet Stop/Running		
3	High Pressure shut down		
4	Low Pressure Shut down		
5	High Oil Pressure Shut down		
6	Low Oil Pressure Shut down		
7	Low Coolant level Shut down		
8	Over speed shut down		
9	Under voltage shut down		
10	Engine failed to start		

c) Switchboards both LV and HV: -

Ser.	Alarm	Received in BMS	Comments
1	ACB or MCCB On/Off Status		
2	ACB/MCCB Trip Status		
3	Battery Charger Failure		
4	Battery Low Voltage Alarm		
5	TVSS Alarm		

d) **All Power Meters, should be able to see all electrical parameters:** -

Ser.	Alarm	Received in BMS	Comments
1	Phase Voltages L1, L2 & L3		
2	Line Voltages L1-L2, L2-L3, L1-L3		
3	Phase Current L1, L2 & L3		
4	Neutral Current		
4	Total Power Kw		
5	Total Power KVA		
6	Total Power KVAr		
7	Total Power Factor		
8	THD L1, L2 and L3		
9	Total Energy kWh		
10	Total Frequency		

e) UPS: -

Ser.	Alarm	Received in BMS	Comments
1	UPS Overload Alarm		
2	UPS on Bypass		
3	UPS Battery Disconnect Alarm		
4	UPS Low Battery Alarm		
5	UPS Battery Discharge		
6	UPS Input Voltages L1, L2 & L3		
7	UPS Input Current L1, L2 & L3		
8	UPS Input Frequency L1, L2 & L3		
9	UPS output Voltages L1, L2 & L3		
10	UPS output Current L1, L2 & L3		
11	UPS output Frequency L1, L2 & L3		
12	UPS output Real Power L1, L2, L3		
13	UPS output KVA L1, L2, L3		
14	UPS bypass voltage L1, L2 & L3		
15	UPS bypass current L1, L2, & L3		
16	UPS bypass Frequency, L1, L2, L3		
17	UPS Battery Current Positive		
18	UPS Battery Current Negative		
19	UPS Battery Time Remain		
20	UPS Battery Voltage Positive		
21	UPS THD L1, L2, L3		

f) STS: -

Ser.	Alarm	Received in BMS	Comments
1	STS Common Alarm		
2	STS on Primary Source		
3	STS on Secondary Source		
4	STS on Bypass		

g) PDU including branch monitoring:

Ser.	Alarm	Received in BMS	Comments
1	PDU Phase Voltages L1, L2 & L3		
2	Line Voltages L1-L2, L2-L3, L1-L3		
3	PDU Phase Current L1, L2 & L3		
4	PDU Neutral Current		
4	PDU Total Power Kw		
5	PDU Total Power KVA		
6	PDU Total Power KVAr		
7	PDU Total Power Factor		
8	PDU Input Breaker Alarm		
9	PDU Total Energy kWh		
10	PDU Total Frequency		
11	PDU Breaker Status		
12	PDU TVSS Status		
13	PDU Primary Temperature Alarm		
14	PDU Secondary Temperature Alarm		

h) ATS:

Ser.	Alarm	Received in BMS	Comments
1	ATS Common Alarm		
2	ATS on Primary Source		
3	ATS on Secondary Source		
4	STS on Bypass		

i) Chilled Water Pump:

Ser.	Alarm	Received in BMS	Comments
1	CHWP Power Status		
2	CHWP E-Stop		
3	Y-D on/Off Status		
4	VFD on/Off Status		
4	Control		
5	Y-D fault Alarm		
6	VFD fault Alarm		
7	DP		
8	VFD		
9	Control		
10	Power kW		

j) Chillers: -

Ser.	Alarm	Received in BMS	Comments
1	Chiller Power Status		
2	Chiller E-Stop		
3	Chiller on/Off Status		
4	Chiller Control		
5	Chiller Common fault Alarm		
6	Chiller Cooling Load		
7	Differential Pressure		
8	VFD %		
9	Control		
10	Power kW		
11	Chiller COP		
12	Auto/Manual		
13	Chiller Inlet Temperature		
14	Chiller Outlet Temperature		
15	Chilled Water flow rate		
16	Chilled water valve Status		
17	Condensing water Valve status		

k) Condensing water Pump

Ser.	Alarm	Received in BMS	Comments
1	CDWP Power Status		
2	CDWP E-Stop		
3	Auto/Manual		
4	Differential Pressure		
5	Control		
6	Power kW		

l) Cooling Tower

Ser.	Alarm	Received in BMS	Comments
1	CT Power Status		
2	CT E-Stop		
3	Y-D on/Off Status		
4	VFD on/Off Status		
4	Control on/off		
5	Y-D fault Alarm		
6	VFD fault Alarm		
7	Auto/Manual		
8	VFD - Hz		
9	Control - hz		
10	Power kW		
11	Vibration Alarm		
12	On/Off Status		
13	Ambient Temperature		
14	CT Inlet Temperature		
15	CT Outlet Temperature		
16	CT Modulating Valve Status		
17	CT Water Flow Rate		

m) CRAC units,

Ser.	Alarm	Received in BMS	Comments
1	CRAC Status		
2	Main Fan Overload		
3	Air Filter Clog		
4	High Temperature alarm		
4	High Humidity Alarm		
5	Fan 1 Fault		
6	Fan 2 Fault		
7	Main Fan Speed		
8	Return Air Humidity		
9	Return Air Temperature		
10	Supply Air Temperature		
11	Fire Trip		

n) AHU/PAU

Ser.	Alarm	Received in BMS	Comments
1	PAU Motorised Fire Damper Status		
2	Pre-Filter Clog Alarm		
3	Bag Filter Clog Alarm		
4	Auto/Manual Mode		
5	Common Fault		
6	Fire Trip		
7	Air Flow Rate		
8	Modulating Valve Status		
9	Supply Air Temperature		
10	Return Air Temperature		
11	Fresh Air Intake Humidity		

o) Data Hall Temperature Sensor

Ser.	Alarm	Received in BMS	Comments
1	Temperature Sensor		
2	Humidity Sensor		

p) AFA panel

Ser.	Alarm	Received in BMS	Comments
1	Common Fault		
2	Detector Fault		
3	Fire		
4	Battery Charger Fault		
5	Battery fault		

q) FM200 System

Ser.	Alarm	Received in BMS	Comments
1	Auto/Manual		
2	Common Fault		
3	Zone Trouble		
4	Fire		
5	Battery Charger Fault		
6	Battery fault		
7	Gas Discharged		

r) **Pre-action System**

Ser.	Alarm	Received in BMS	Comments
1	Common Alarm		
2	Compressor Cut-in alarm		
3	Zone 1 smoke detector alarm		
4	Zone 2 smoke detector alarm		
5	Zone 1 Low Pressure Alarm		
6	Zone 2 Low Pressure Alarm		
7	AC Power Loss Alarm		
8	Water Discharge Aalrm		

s) **Sprinkler System**

Ser.	Alarm	Received in BMS	Comments
1	Sprinkler Activated		
2	Sprinkler Pump Running		
3	Jocky Pump Running		

t) **VESDA System**

Ser.	Alarm	Received in BMS	Comments
1	Common Alarm		
2	Alert alarm		
3	Fire 1 Alarm		
4	Fire 2 Alarm		
5	Fire 3 Alarm		
6	Filter Clog alarm		
7	AC Power Fault		
8	Battery fault		

u) Sump Pumps

Ser.	Alarm	Received in BMS	Comments
1	Auto/Manual		
2	On/Off Status		
3	High Level Alarm		
4	Low level Alarm		
5	Overload Trip		

v) Water Leakage Detection System

Ser.	Alarm	Received in BMS	Comments
1	Common Alarm		
2	Cable Break alarm		
3	Water Leakage at x metres		

w) Lighting control system by zones

Ser.	Alarm	Received in BMS	Comments
1	On/Off Remotely		
2	Bypass Switch On		

CHAPTER 7

Appendices

Appendix A: IST Method Statement
Appendix B: Heat Load Test Method Statement.

APPENDIX A

IST Method Statement Sample

CONTENTS

1. Introduction 97
2. Pre-requisites 99
3. Program of Events 102
4. List and Location of Participants 103
5. Instrumentation Record Sheets 104
6. Cause and Effect Scenarios 105

   6.1   TEST 1 - Failure of TX-D6 105
   6.2   TEST 2 - Failure of TX-D7 108
   6.3   TEST 3 - Failure of TX-D8 109

1

Introduction

The Hill Top Data Centre International Limited (HTDCIL) is located on a 23,807 m² site at Mind Space, Hyderabad, India

There will be 8 data centre halls in the first phase (Day 1) – four on the third floor and the other 4 halls on the fourth floor. This Integrated Systems Test (IST) method statement is prepared for the infrastructure and facilities in the first phase only. The maximum design IT load capacity for phase 1 is 1443.08kW (35.68+315+286+806.4) (3/F Data Halls 636.68+ 806.4 from 4/F Data Halls).

The Integrated Systems Test is the accepted industry-standard procedure to prove the systems installed in the facility will provide the Client with the resilience they require for a given equipment failure scenario.

This IST covers all the critical equipment in the premises. The objective is to demonstrate the performance of all the systems serving the premises in a controlled environment under which the equipment can have simulated fault or failure conditions introduced to ensure the equipment functions and systems react in a manner to which they are designed.

There should be no intervention of the system operation by third parties unless a serious situation develops or during the resetting of the system as set out in the test script after completion of each failure scenario.

This Integrated Systems Test is scheduled for **May 12-20, 2018**. This will be depended on approval of Construction Programme M2 – M5.

The tests shall be orchestrated from a command centre set up in G/F **NOC**. The command centre will be in radio communication with all parties involved in the testing for the relaying of information back to the test coordinator.

2

Pre-requisites

The maximum design day 1, IT load shall be connected in the data halls for the IST.

The following systems and the interfaces between them must be complete and verified by Data Centre Testing Consultancy Asia Limited prior to the IST: -

Electrical Systems

Electrical Power and Lighting, Emergency Generators and Switchboards servicing the facility, with dual UPS source of power, consisting of the following components and sub-systems.

- 12 nos. 2000kVA transformers (by utility provider) located at G/F
- 1 no 1500kVA transformers located at G/F
- 7 nos. 2250kVA + 1 no. 800kVA emergency generators located on R/F and associated fuel supply equipment
- 8 nos. generator local switchboards located on R/F
- 8 nos. generator changeover switchboards located on R/F
- 6 nos. main incoming low voltage switchboards located on G/F
- 10 nos. MCC Switchboard c/w ATS located on R/F
- 190 Main Distribution Boards
- 8 nos. UPS switchboards located on 3/F
- 8 nos. UPS switchboards located on 4/F
- 4 nos. 500kVA UPS located on 3/F
- 6 nos. 500kVA UPS located on 4/F
- 12 nos. 400A STS located on 3/F
- 12 nos. 400A STS located on 4/F

- 12 nos. 225kVA PDU located on 3/F
- 12 nos. 225kVA PDU located on 4/F
- 4 nos. 70kw DCPSS unit's c/w PDU units
- Automatic Transfer Switches (ATS)

Fire Services Systems

The Fire Services Systems servicing the facility consisting of the following components and sub-systems.

- AFA System
- Gas Suppression System
- Pre-Action Sprinkler System
- Wet Sprinkler System
- 2 nos. 55kW Sprinkler Pumps c/w 2 nos. 3kW Jockey Pump
- 2 nos. 37kW Fixed Pumps c/w 3kW Sprinkler Jockey Pump
- Aspirating Smoke Detection System
- Smoke and heat detectors
- Hose Reels
- Hydrants

Mechanical Systems

The Air Conditioning and Mechanical Ventilation Systems servicing the facility consists of the following components and sub-systems.

- 5 nos. 450TR water-cooled Chiller located on R/F
- 5 nos. 45kW Cooling Towers located on R/F
- 5 nos. 45kW Chilled water pumps located on R/F
- 5 nos. 50kW Condensate water pumps located on R/F
- 5 nos. 2.2kW Chemical dozing pumps located on R/F
- 5 nos. 5.5kW Separator pumps for cooling tower on R/F
- 42 nos. 100kW CRAC units
- 4 nos. 94kW CRAC units
- 4 nos. 50kW CRAC units

- 4 nos. 33kW CRAC Units
- 38 nos. Fan Coil Units
- 69 Exhaust Air Fans
- 4 Fresh Air Fans
- Chillers Control System (CCS)
- 6 nos. Air Handling Units (AHU)
- 9 nos. Primary Air Handling Units (PAU)
- 2 nos. Heat Exchanger, c/w 2 nos. Circulation pumps
- 8 nos. Split Type Air-conditioning Units
- 152 VAV boxes
- 2 nos. Make-up Water pumps
- Water Leak Detection System

Building Management System (BMS)

The Building Management System servicing the facility consisting of the following components and sub-systems.

- DDC panels, Building Network Controllers, High Level Interface Units, patch panels, Edge switches, Core switches and printers
- Digital Multifunction Power Meters (CVM) and Digital Power Analysers (DPA)

Lift

- 2 nos. Passenger Lifts and 2 nos. Goods Lift

3

Program of Events

The events will be consolidated prior to the date via a succession of pre-event meeting whereby an hour by hour programme will be developed but the testing is scheduled to be conducted over one day.

<u>Programme Summary</u>

- Roll call / Pre-Test briefing
- Radio and instrument check
- Personnel location check
- Normal Condition clarification
- Testing commences
- Collate data and debrief
- Conclusion of IST

A more detailed programme is included in Appendix A.

4

List and Location of Participants

The following list indicates the minimum number of participants required for the Integrated Systems Test. The expected locations of the main parties involved in the IST are shown below: -

No.	Name	Company	Role for IST	Location
1.				
2.				
3.				
4.				
5.				
6.				
7.				
8.				
9.				
10.				
11.				
12.				
13.				
14.				
15.				
16.				
17.				
18.				
19.				
20.				
21.				
22.				
23.				
24.				
25.				

5

Instrumentation Record Sheets

Company	Instrument	Calibration Cert.	Location
	Walkie-talkies		
	Clamp meters		
	Multi-meters		
	Heat loads		
	Sling psychrometer		
	Digital thermometers		
	Power Analysers		

6

Cause and Effect Scenarios

This section provides an overall summary of the equipment that is effected by the failure scenarios and the expected reactions.

Normal Conditions

These are the normal operating conditions of all systems before each test. After each test, the systems will be restored to their normal operating conditions before the commencement of the next test.

6.1   TEST 1 - Failure of TX-D6

Simulation: -

Open 2500A incoming ACB at LV switchboard LVSB/SR02/02 in LV Switch Room 02, at G/F

Effect

Stage 1

- Loss of power to Electrical Vehicle Chargers, Landscape L&P at G/F
- Loss of power to all General L&P at 1/F
- Loss of power to all Public L&P at 1/F and 2/F
- Loss of power to all Public L&P at 3/F- 5/F
- Loss of power to all Public L&P at R/F
- Loss of power to all AHU, PAU & EAF fed from MB-BB-AC-1F-01N
- Loss of power to all VAV & HTR fed from MB-BB-AC-1F-01N

- Loss of power to MB-BB-PD-GF-01N
- Loss of power to Kitchen & Canteen fed from MB-FB-01N
- Loss of power to 2/F Office area fed from MB-OP-2F-01N
- Loss of power to Security Guard House fed from DB-BB-GL-GF-01E
- Loss of power to General Power G/F MB-BB-GL-GF-01E
- Loss of power to 1/F and 2/F essential L&P MB-BB-PP-1F-01E
- Loss of power to CRAC Boards 3/F & 4/F fed from MB-BB-PP-3F-01E
- Loss of power to Essential Public L&P at 5/F fed from MB-BB-PP-4F-01E
- Loss of power to Essential Public L&P at R/F & UR/F fed from MB-BB-PP-RF-01E
- Loss of power to Passenger Lifts # 1 & 2 fed from MB-BB-5F-01E1 & E2
- Loss of power to Cargo Lift & MTV Lift fed from MB-BB-LT-RF-01E and MB-BB-LT-5F-02E.
- Loss of power to all F.S pumps fed from LMCP-FS-L1-01E & LMCP-FS-L1-02E.
- Loss of power to 300kVAR capacitor bank at G/F
- Loss of power to 2/F Kitchen & Canteen fed from MB-FB-2F-01E
- Genset G-FS starts

Stage 2

- Loss of power to Electrical Vehicle Chargers, Landscape L&P at G/F
- Loss of power to all General L&P at 1/F
- Loss of power to all Public L&P at 1/F and 2/F
- Loss of power to all Public L&P at 3/F- 5/F
- Loss of power to all Public L&P at R/F
- Loss of power to all AHU, PAU & EAF fed from MB-BB-AC-1F-01N
- Loss of power to all VAV & HTR fed from MB-BB-AC-1F-01N
- Loss of power to MB-BB-PD-GF-01N

- Loss of power to Kitchen & Canteen fed from MB-FB-01N
- Loss of power to 2/F Office area fed from MB-OP-2F-01N
- Genset G-FS Running and feeds power to all essential loads
- Resume power to Security Guard House fed from DB-BB-GL-GF-01E
- Resume power to General Power G/F MB-BB-GL-GF-01E
- Resume power to 1/F and 2/F essential L&P MB-BB-PP-1F-01E
- Resume power to CRAC Boards 3/F & 4/F fed from MB-BB-PP-3F-01E
- Resume power to Essential Public L&P at 5/F fed from MB-BB-PP-4F-01E
- Resume power to Essential Public L&P at R/F & UR/F fed from MB-BB-PP-RF-01E
- Resume power to Passenger Lifts # 1 & 2 fed from MB-BB-5F-01E1 & E2
- Resume power to Cargo Lift & MTV Lift fed from MB-BB-LT-RF-01E and MB-BB-LT-5F-02E.
- Resume power to all F.S pumps fed from LMCP-FS-L1-01E & LMCP-FS-L1-02E.
- Resume power to 300kVAR capacitor bank at G/F
- Resume power to 2/F Kitchen & Canteen fed from MB-FB-2F-01E

Restore of TX-D6

Simulation: -

Close 2500A incoming ACB at LV switchboard LVSB/SR02/02 in LV Switch Room 02, at G/F

Effect

- Resumption of Normal power to all areas after time delay of 5 seconds
- **Generator G-FS starts 5-minutes cool-down period and then stops**

Reset to Normal Condition

6.2 TEST 2 - Failure of TX-D7

Simulation: -

Open 3200A outgoing ACB at LV switchboard LVSB/SR03/01 in Switch Room 03, G/F

Effect

Stage 1

- Loss of main power to switchboard LVSB/4-1-B1/01 in UPS Room 01
- Loss of power to LVSB/4-1-B1/02
- Loss of power to LVSB/4-1-B1/03
- Genset G01 starts after delay timers start (2 seconds)

Stage 2

- Genset G01 running on load
- Genset G01 feeds power to switchboard LVSB/4-1-B1/01 in UPS Room 01
- LVSB/4-1-B1/02 on Genset supply
- Genset power to LVSB/4-1-B1/03

Restore of TX-D7

Simulation: -

Close TX-D2 3200A outgoing ACB at LV switchboard LVSB/SR03/01 in Switch Room 03, G/F

Effect

- Resumption of main B supply to switchboard LVSB/4-1-B1/01
- Delay timer of ATS at switchboard LVSB/4-1-B1/01 starts (5 Secs)
- After 5 Secs, 3200A ACB M1 (NO) of ATS at switchboard LVSB/4-1-B1/01 opens
- 3200A ACB M2 (NC) of ATS at switchboard LVSB/4-1-B1/01 closes
- **Generator G-01 starts 5-minutes cool-down period and then stops**

Reset to Normal Condition

6.3 TEST 3 - Failure of TX-D8

Simulation: -

Open TX-D8 3200A outgoing ACB at LV switchboard LVSB/SR03/02 in Switch Room 03 on G/F

Effect:

Stage 1

- Loss of main power to switchboard LVSB/4-2-B1/01 in UPS Room 04
- UPS-4-2-B1, UPS-4-2-B2 and UPS-4-2-B3 on Battery
- LVSB/4-2-B1/02 on UPS
- STS-4-3-B1, STS-4-3-B2, STS-4-3-B3 on preferred source
- PDU-4-3-B1, PDU-4-3-B2, PDU-4-3-B3 on UPS load
- STS-4-4-B1, STS-4-4-B2, STS-4-4-B3 on preferred source
- PDU-4-4-B1, PDU-4-4-B2, PDU-4-4-B3 on UPS load
- Loss of power to LVSB/4-2-B1/03
- Loss of power to ISO-CR05-4F-B1

- Loss of power to ISO-CR06-4F-B1
- Loss of power to ISO-CR07-4F-B1
- Loss of power to ISO-CR08-4F-B1
- All CRACs fed from source B in Data Halls 3 & 4 stops
- Genset G02 starts after delay timers start (2 seconds)

Stage 2

- Genset G02 running on load
- Genset G02 feeds power to switchboard LVSB/4-2-B1/01 in UPS Room 04
- UPS-4-2-B1, UPS-4-2-B2 and UPS-4-2-B3 on Rectifier
- LVSB/4-2-B1/02 on Genset supply
- STS-4-3-B1, STS-4-3-B2, STS-4-3-B3 on preferred source
- PDU-4-3-B1, PDU-4-3-B2, PDU-4-3-B3 on Genset load
- STS-4-4-B1, STS-4-4-B2, STS-4-4-B3 on preferred source
- PDU-4-4-B1, PDU-4-4-B2, PDU-4-4-B3 on Genset load
- Genset power to LVSB/4-2-B1/03
- Genset power to ISO-CR05-4F-B1
- Genset power to ISO-CR06-4F-B1
- Genset power to ISO-CR07-4F-B1
- Gense power to ISO-CR08-4F-B1
- All CRACs fed from source B in Data Halls 3 & 4 starts after time delay

Restore of TX-D8

Simulation: -

Close TX-D8 3200A outgoing ACB at LV switchboard LVSB/SR03/02 in LV Switch Room 03 on G/F

Effect

- Resumption of mains B supply for switchboard LVSB/4-2-B1/01
- Delay timer of ATS at switchboard LVSB/4-2-B1/01 starts (5 Secs)
- After 5 Secs, 3200A ACB M1 (NO) of ATS at switchboard LVSB/4-2-B1/01 opens
- 3200A ACB M2 (NC) of ATS at switchboard LVSB/4-2-B1/01 closes
- **Generator G02 starts 5-minutes cool-down period and then stops**

Reset to Normal Condition

APPENDIX B

HLT Method Statement Sample

CONTENTS

1. Introduction 117
2. Pre-requisites Conditions 119
3. Program of Events 121
4. List of Participants 122
5. Location of Participants 123
6. Instrumentation 124
7. Heat Load Test Scenarios 125

   Test 0 – Normal Operation Check – No load condition 126
   Test 1 – Heat Load Test at 25% simulated I.T. load 127
   Test 2 – Heat Load Test at 50% simulated I.T. load 128
   Test 3 – Heat Load Test at 75% simulated I.T. load 129
   Test 4 – Heat Load Test at 100% simulated I.T. load 130
   Test 5 – Heat Load Test at 100% load with adjacent CRAC units of two groups failed (CRAC Units 4 & 5) 131
   Test 6 – Heat Load Test at 100% load with adjacent CRAC unit of two failed (CRAC Units 10 & 11) 132
   Test 7 – Fire Trip Test shutting down all CRAC at 100% heat load 133
   Test 8 – Test UPS Room CRAC Failure at 100% I.T. Load 134
   Test 9 – Test Battery Room FCU Failure at 100% I.T. Load 136
   Test 10 – Heat Load Test at 100% load with Riser (path A) Shut Off 138

Test 11 – Heat Load Test at 100% load with Riser (path B) Shut Off..139
Test 12 – CRAC Communication Failure at 100% load.......140

Appendix 1 – Heat Load Test Schedule 141
Appendix 2 – Data Logger Layout..145

1

Introduction

The Data Centre located at 2 Hill Top Road Bangalore is built to serve for JP Cross's Data Centre needs.

The Heat Load Test is the process to prove the Data Centre air conditioning system design and installation is capable to sustain against any system failures and change-over statues.

The Air Conditioning Cooling Systems is the chilled water system supplied by 2 sets of chillers (Ch-8 & Ch-9) on the roof which provides chilled water via 2 sets of riser routing to 2/F Data Centre area.

The main chilled water header is formed in ring circuit in Data Centre for supplying up to CRAC units.

There is one (1) Data Centre at 2/F being tested under this heat load test. There are 2 groups of CRAC with total 12 sets CRAC units. Each group is on hot standby to ensure reliable cooling to Data Centre environment.

Location	Data Centre
Area	1000 sq.m
Anticipate number of racks space	275
Total IT heat load	962.5 kW
Design cold aisle temperature	21 - 25 oC

Design Criteria for Data Centre

The following test procedure has been prepared as a guideline for the heat load performance testing of both the normal and emergency operating conditions of the Data Centre.

The heat load test is to demonstrate the performance of the cooling systems with an artificial load (in full and partial load conditions) and to monitor/record the effects of simulated failures and changeover of system operation.

Prior to undertaking the test/s it is essential that all systems associated with the areas under test have been thoroughly tested, commissioned and verified to be operating in accordance with the design criteria.

2

Pre-requisites Conditions

This Heat Load Test is focused on the air conditioning system operation for Data Centre areas.

The following systems must be completed and verified by customer representatives and Data Centre Testing and Consultancy Asia Limited prior to the Test: -

1. Electrical Systems

All electrical system tests have been completed and normal powers are available for the Heat Load Test.

The dummy load fan heaters shall be connected evenly as per design criteria to simulate the heat load distribution.

2. Mechanical Systems

The MVAC Systems servicing the facility consisting of the following components and sub-systems.

- CRAC units are in normal service for Data Centre area.
- Chilled water distribution system is balanced
- Condensate drain system is in normal condition.
- Chiller control system is in normal conditions.
- CRAC units testing is completed.
- All perforated floor grilles air balancing test is completed

Fire Services Systems

The Fire Services Systems is preferable completed and tested with AFA system in normal operation. The FM 200 system was applied at the LV switch Room and Pre-action system applied at data center. All Fire Services System shall be disconnected during Heat Load Test to avoid any false alarm.

Portable fire extinguishers shall be provided and evenly allocated inside the Data Centre for emergency operation in case any smoke and fire caused by the dummy load.

Building Management System (BMS)

The Building Automation Systems servicing the facility consist of the following components and sub-systems.

- BMS head end computer interface.
- DDC panels located near the major equipment and signal sampling area.
- ALL interfaces between the systems must have been verified.
- BMS system test shall be completed.
- Data Centre temperature profile shall be monitored during Heat Load Test.
- CRAC operation status shall be monitored during Heat Load Test.
- Data logger is set at a level of 1.5m above raised floor for temperature and humidity sensing & record.
- Data loggers are installed and distributed for recording full temperature profile during Heat Load Test. (Appendix - 2)

3

Program of Events

The events will be consolidated prior to the date via a succession of pre-event meeting whereby an hour by hour program will be developed but the testing is scheduled to be conducted over two days.

- Registration for duty reporting
- Roll call / Pre-Test briefing
- Walkie Talkie Radio check
- Instrumentation check
- Personnel location check
- Test commencement announcement
- Normal Condition clarification
- Testing commences

Appendix A shows the proposed time table

4

List of Participants

No.	Name	Company	Title	Location
1.				
2.				
3.				
4.				
5.				
6.				
7.				
8.				
9.				
10.				
11.				
12.				
13.				
14.				
15.				
16.				
17.				
18.				

5

Location of Participants

Early plan for the participants and way of communication shall ensure the Heat Load Test smooth run down to provide reasonable simulation and verification.

Observers are limited to roam around the different areas as the heat load test shall control the room integrity and provide minimum interference to temperature record.

6

Instrumentation

Company	Instrument	Calibration Cert.	Location

7

Heat Load Test Scenarios

This section elaborates the scenarios for heat load test verification on the Data Centre air conditioning system performance.

The Data Centre requirements are listed as below, the set points are proposed for testing and may require adjustment during testing.

Location	Data Centre
Area (White Space)	1000 sq m
Anticipate number of racks space	275
Total heat load (To be equally distributed between UPS-A & -B system)	962.5 kW
Dummy load (Fan Heater) Quantity (To be adjusted accordingly if heat load is different from listed)	535 nos. @ 1.8kW
Design cold aisle temperature	21 - 25 oC
CRAC supply air temperature set point	16 – 18 oC

The following scenarios tests are try to simulate most of the possible operation status, change over and fault events during the Data Centre operation. The temperature profile of the cold aisle will be recorded by data loggers while the BMS system will monitor the overview conditions for each test. The BMS will trend and record the supply and return temperatures of all CRAC units.

Test 0 – Normal Operation Check – No load condition

Check to ensure ALL systems are in their normal operating conditions and are ready to accept the failures proposed.

1. Chillers and pumps are in normal operation.
2. Chilled water leaving temperature set at 10oC
3. CRAC units are on hot standby at each group
4. Condensate pump & drain in normal operation
5. Dummy heat load – fan heaters and hair dryers are connected to full capacity loading requirement
6. Data loggers are set properly at cold aisles and hot aisles
7. Data Centre temperature sensors are monitoring properly
8. Check Data Centre at no load condition; run Data Centre for 30mins and monitor the Data Centre temperature until in stable condition.

Location	Data Centre
Area	1000 sq.m
Anticipate number of racks space	275
Total design heat load (To be equally distributed between UPS-A & -B system)	962.5 kW
Dummy load Quantity in operation	No load
Design cold aisle temperature	21 - 25 oC
CRAC unit in operation	12

Test 1 – Heat Load Test at 25% simulated I.T. load

Test Objective:

This test will be carried out at 25% design heat load to simulate the full I.T. load condition. Also take measurement reading for PUE value.

1. Set 25% dummy load heater.
2. Monitoring Data Centre temperature rise and CRAC ramp up cooling in operation.
3. Check BMS temperature profile to be stable in response to light heat load for 1 hour
4. Upon 30mins running, if room temperature still showing up trend and not stabilized yet, continuous to keep Data Centre running until the Data Centre room temperature stabilized.

Location	Data Centre
Area	1000 sq.m
Anticipate number of racks space	275
Total design heat load	962.5 KW
Dummy load quantity in operation (To be equally distributed between UPS-A & -B system)	241 KW (25%)
Design cold aisle temperature	21 - 25 oC
CRAC units in operation	12

Test 2 – Heat Load Test at 50% simulated I.T. load

Test Objective:

This test will be carried out at 50% design heat load to simulate the full I.T. load condition. Also take measurement reading for PUE value.

1. Set 50% dummy load heater.
2. Monitoring Data Centre temperature rise and CRAC ramp up cooling in operation.
3. Check Data Centre BMS temperature profile to be stable in response to light heat load for 1 hour
4. Upon 30mins running, if room temperature still showing up trend and not stabilized yet, continuous to keep Data Centre running until Data Centre room temperature stabilized.

Location	Data Centre
Area	1000 sq.m
Anticipate number of racks space	275
Total design heat load	962.5 KW
Dummy load Quantity in operation (To be equally distributed between UPS-A & -B system)	481 KW (50%)
Design cold aisle temperature	21 - 25 oC
CRAC unit in operation	12

Test 3 – Heat Load Test at 75% simulated I.T. load

Test Objective:

This test will be carried out at 75% design heat load to simulate the full I.T. load condition. Also take measurement reading for PUE value.

1. Set 75% dummy load heater.
2. Monitoring Data Centre temperature rise and CRAC ramp up cooling in operation.
3. Check BMS temperature profile to be stable in response to light heat load for 1 hour
4. Upon 30mins running, if room temperature still showing up trend and not stabilized yet, continuous to keep Data Centre running until Data Centre e room temperature stabilized.

Location	Data Centre
Area	1000 sq.m
Anticipate number of racks space	275
Total design heat load	962.5 KW
Dummy load quantity in operation (To be equally distributed between UPS-A & -B system)	722 KW (75%)
Design cold aisle temperature	21 - 25 oC
CRAC unit in operation	12

Test 4 – Heat Load Test at 100% simulated I.T. load

Test Objective:

This test will be carried out at 100% design heat load to simulate the full I.T. load condition. Also take measurement reading for PUE value.

1. Set 100% dummy load heater.
2. Monitoring Data Centre temperature rise and CRAC ramp up cooling in operation.
3. Check BMS temperature profile to be stable in response to light heat load for 1 hour.
4. Upon 1hour running, if room temperature still showing up trend and not stabilized yet, continuous to keep Data Centre running until Data Centre room temperature stabilized.

Location	Data Centre
Area	1000 sq.m
Anticipate number of racks space	275
Total design heat load	962.5 KW
Dummy load quantity in operation (To be equally distributed between UPS-A & -B system)	962.5 KW (100%)
Design cold aisle temperature	21 - 25 oC
CRAC unit in operation	12

Test 5 – Heat Load Test at 100% load with adjacent CRAC units of two groups failed (CRAC Units 4 & 5)

Test Objective:

This test will be carried out at 100% heat load to simulate the temperature profile for when there are 2 sets of CRAC units (Nos 4 and 5) in adjacent locations but in different groups

1. At 100% heat load condition, simulate fan failure of CRAC units 4 & 5 while the rest of the CRAC units are running in normal mode.
2. Monitor Data Centre temperature for any fluctuation or hot spot.
3. Check BMS temperature profile to be stable in response to both CRAC stop running adjacent to each other. Monitor heat load for 30mins.
4. Upon 30mins running, if room temperature still showing up trend and not stabilized yet, continue to monitor Data Centre until Data Centre room temperature stabilized.

Location	Data Centre
Area	1000 sq.m
Anticipate number of racks space	275
Total design heat load	962.5 KW
Dummy load quantity in operation (To be equally distributed between UPS-A & -B system)	962.5 KW (100%)
Design cold aisle temperature	21 - 25 oC
CRAC unit in operation	10

Test 6 – Heat Load Test at 100% load with adjacent CRAC unit of two failed (CRAC Units 10 & 11)

Test Objective:

This test will be carried out at 100% heat load to simulate the temperature profile for when there are 2 sets of CRAC units (nos 10 & 11) in adjacent locations but in different groups

1. At 100% heat load condition, simulate fan failure of CRAC units 10 & 11 while the rest of the CRAC units are running in normal mode.
2. Monitor Data Centre temperature for any fluctuation or hot spot.
3. Check BMS temperature profile to be stable in response to both CRAC stop running adjacent to each other. Monitor heat load for 30mins.
4. Upon 30mins running, if room temperature still showing up trend and not stabilized yet, continue to monitor Data Centre until Data Centre room temperature stabilized.

Location	Data Centre
Area	1000 sq.m
Anticipate number of racks space	275
Total design heat load	962.5 KW
Dummy load quantity in operation (To be equally distributed between UPS-A & -B system)	962.5 KW (100%)
Design cold aisle temperature	21 - 25 oC
CRAC unit in operation	10

Test 7 – Fire Trip Test shutting down all CRAC at 100% heat load

Test Objective:

This test will be carried out at 100% heat load to simulate the temperature profile for when there is a fire trip shutting down all CRAC units. The reference temperature sensor in the Data Centre shall be taken as that which first reaches 35 deg C and sends an alarm to BMS.

Two reference temperature sensors are required. The time taken from the moment the CRAC units are tripped will be recorded.

Two technicians will be assigned to reset the fire trip signal and CRAC units when Data Centre temperature reached 35 deg C.

1. At 100% heat load condition, simulate a fire trip event which shut down all the CRAC operation.
2. Monitoring Data Centre temperature rise up to 35oC as monitored by BMS.
3. Record BMS temperature rise profile and timing.
4. Restart all CRAC units to cool down the Data Centre.
5. Keep checking on BMS temperature profile and timing until the Data Centre e room temperature is stabilized.

Location	Data Centre
Area	1000 sq.m
Anticipate number of racks space	275
Total design heat load	962.5 KW
Dummy load quantity in operation (To be equally distributed between UPS-A & -B system)	962.5 KW (100%)
Design cold aisle temperature	21 - 25 oC
CRAC unit in operation	12

Test 8 – Test UPS Room CRAC Failure at 100% I.T. Load

Test Objective:

This test will be carried out with 100% heat load for UPS room 2-H1, 2-J1 and 2-K1 to simulate a fan failure alarm at a CRAC unit at UPS room. There are 2 sets of CRAC unit in hot standby configuration in each of the UPS Rooms. The design return temperature for the UPS rooms is 28 degrees Celsius

Test – 8a: CRAC-49 & CRAC-50 for UPS room 2-H1

1. At 100% Data Centre heat load condition, UPS is supporting full Data Centre design load condition. Simulate the CRAC-49 fan failure which causes CRAC-50 to take up the entire load in UPS room 2-H1.
2. Monitor UPS room temperature for any fluctuation or high temperature alarm.
3. Check BMS temperature profile to be stable in response to Change-over and monitor the temperature profile for 15mins.
4. Upon 15mins running, if room temperature still showing up trend and not stabilized yet, continues to keep test going until UPS room temperature has stabilized.

Test – 8b: CRAC-51 & CRAC-52 for UPS room 2-J1

1. At 100% Data Centre heat load condition, UPS is supporting full design load condition. Simulate the CRAC-51 fan failure which causes CRAC-52 to take up the entire load in UPS room 2-J1.
2. Monitor UPS room temperature for any fluctuation or high temperature alarm.
3. Check BMS temperature profile to be stable in response to Change-over and monitor the temperature profile for 15mins.

4. Upon 15mins running, if room temperature still showing up trend and not stabilized yet, continues to keep test going until UPS room temperature has stabilized.

Test – 8c: CRAC-53 & CRAC-54 for UPS room 2-K1

1. At 100% Data Centre heat load condition, UPS is supporting full design load condition. Simulate the CRAC-53 fan failure which causes CRAC-54 to take up the entire load in UPS room 2-K1
2. Monitor UPS room temperature for any fluctuation or high temperature alarm.
3. Check BMS temperature profile to be stable in response to Change-over and monitor the temperature profile for 15mins.
4. Upon 15mins running, if room temperature still showing up trend and not stabilized yet, continues to keep test going until UPS room temperature has stabilized.

Location	Data Centre
Area	1000 sq.m
Anticipate number of racks space	275
Total design heat load	962.5 KW
Dummy load quantity in operation (To be equally distributed between UPS-A & -B system)	962.5 KW (100%)
Design cold aisle temperature	21 - 25 oC
CRAC unit in operation	12

Test 9 – Test Battery Room FCU Failure at 100% I.T. Load

Test Objective:

This test will be carried out with 100% heat load for Battery Rooms 2-1, 2-2 and 2-3 to simulate an alarm at a FCU at Battery room. There are 2 sets of FCU in hot standby configuration in each of the Battery Rooms.

Test – 9a: FCU 48 & FCU 49 for Battery Room 2-1

1. At 100% Data Centre heat load condition, simulate FCU 48 failure which causes FCU 49 to take up the entire load in Battery Room 2-1
2. Monitor Battery room temperature for any fluctuation or high temperature alarm.
3. Check BMS temperature profile to be stable in response to Change-over and monitor the temperature profile for 15mins.
4. Upon 15mins running, if room temperature still showing up trend and not stabilized yet, continues to keep test going until Battery room temperature has stabilized.

Test – 9b: FCU 50 & FCU 51 for Battery Room 2-2

1. At 100% Data Centre heat load condition, simulate FCU 50 failure which causes FCU 51 to take up the entire load in Battery Room 2-2.
5. Monitor Battery room temperature for any fluctuation or high temperature alarm.
6. Check BMS temperature profile to be stable in response to Change-over and monitor the temperature profile for 15mins.
7. Upon 15mins running, if room temperature still showing up trend and not stabilized yet, continues to keep test going until Battery room temperature has stabilized.

Test – 9c: FCU 52 & FCU 53 for Battery Room 2-3

1. At 100% Data Centre heat load condition, simulate FCU 52 failure which causes FCU 53 to take up the entire load in Battery Room 2-3.
2. Monitor Battery room temperature for any fluctuation or high temperature alarm.
3. Check BMS temperature profile to be stable in response to Change-over and monitor the temperature profile for 15mins.
4. Upon 15mins running, if room temperature still showing up trend and not stabilized yet, continues to keep test going until Battery room temperature has stabilized.

Location	Data Centre
Area	1000 sq.m
Anticipate number of racks space	275
Total design heat load	962.5 KW
Dummy load quantity in operation (To be equally distributed between UPS-A & -B system)	962.5 KW (100%)
Design cold aisle temperature	21 - 25 oC
CRAC unit in operation	12

Test 10 – Heat Load Test at 100% load with Riser (path A) Shut Off

Test Objective:

This test will be carried out at 100% heat load to simulate the temperature profile for when one of the chilled water riser supply fail.

The temperature profile record also monitors the trend report when the main zone valve of individual DC shut off to simulate the riser failure and the time for opening the Data Centre zone valves to provide chilled water flow from the other end to the whole Data Centre CRAC supply.

1. At 100% heat load condition, simulate chilled water riser failure by shutting off the riser supply and return valves at incoming section for Path A.
2. Then open subsidiary zone valves in ring header inside the data center to provide the other feed of chilled water circulation to all Data Centre CRAC units.
3. Monitor Data Centre temperature profile by BMS for 1 hour.
4. Record BMS temperature rise profile and timing for 1 hour.
5. Monitor all CRAC units in hot standby mode to cool down the Data Centre
6. Keep checking on BMS temperature profile and timing until the Data Centre room temperature is stabilized.

Location	Data Centre
Area	1000 sq.m
Anticipate number of racks space	275
Total design heat load	962.5 KW
Dummy load quantity in operation (To be equally distributed between UPS-A & -B system)	962.5 KW (100%)
Design cold aisle temperature	21 - 25 oC
CRAC unit in operation	12

Test 11 – Heat Load Test at 100% load with Riser (path B) Shut Off

Test Objective:

This test will be carried out at 100% heat load to simulate the temperature profile for when one of the chilled water riser supply fail.

The temperature profile record also monitors the trend report when the main zone valve of individual DC shut off to simulate the riser failure and the time for opening the Data Centre zone valves to provide chilled water flow from the other end to the whole Data Centre CRAC supply.

1. At 100% heat load condition, simulate chilled water riser failure by shutting off the riser supply and return valves at incoming section for Path B
2. Then open subsidiary zone valves in ring header inside the data centre to provide the other feed of chilled water circulation to all Data Centre CRAC units.
3. Monitor Data Centre temperature profile by BMS for 1 hour.
4. Record BMS temperature rise profile and timing for 1 hour.
5. Monitor all CRAC units in hot standby mode to cool down the Data Centre
6. Keep checking on BMS temperature profile and timing until the Data Centre room temperature is stabilized.

Location	Data Centre
Area	1000 sq.m
Anticipate number of racks space	275
Total design heat load	962.5 KW
Dummy load quantity in operation (To be equally distributed between UPS-A & -B system)	962.5 KW (100%)
Design cold aisle temperature	21 - 25 oC
CRAC unit in operation	12

Test 12 – CRAC Communication Failure at 100% load

Test Objective:

This test will be carried out at 100% heat load to simulate the temperature profile when there is a CRAC communication failure

During the test, the CRAC communication cable shall be disconnected from the master CRAC of both groups.

1. At 100% heat load condition, simulate CRAC communication failure by disconnecting the communication cable from the master CRAC units of the two groups.
2. All CRAC units shall operate normally.
3. Monitor Data Centre temperature profile by BMS for 1 hour.
4. Record BMS temperature rise profile and timing for 1 hour.
5. Monitor all CRAC units in hot standby mode to cool down the Data Centre
6. Keep checking on BMS temperature profile and timing until the Data Centre room temperature is stabilized.

Location	Data Centre
Area	1000 sq.m
Anticipate number of racks space	275
Total design heat load	962.5 KW
Dummy load quantity in operation (To be equally distributed between UPS-A & -B system)	962.5 KW (100%)
Design cold aisle temperature	21 - 25 oC
CRAC unit in operation	12

APPENDIX 1

Heat Load Test Schedule

The Heat Load Test is planned to be carried out in 2 days.

Date	From	To	Tasks	Remarks
			Day-1 Heat Load Test	
1	9:00	9:15	Registration for duty reporting	
2	9:00	9:30	Technicians set up first test condition.	
3	9:15	9:45	Roll call / Pre-Test briefing	
4	9:45	10:00	Walkie Talkie Radio check	
5	10:00	10:05	Instrumentation check	
6	10:05	10:15	Personnel location check	
7	10:15	10:20	Test commencement announcement	
8	10:20	10:45	Normal Condition data centre operation check	
9			**Testing commences**	
10	10:45	11:15	Test – 0 Normal Operation Check (No load condition)	
11	11:15	11:45	Test – 1 Heat Load Test at 25% load	

Item	From	To	Tasks	Remarks
12	11:45	12:15	Test – 2 Heat Load Test at 50% load	
13	12:15	12:45	Test – 3 Heat Load Test at 75% load	
14	12:45	13:15	Test – 4 Heat Load Test at 100% load	
15	13:15	14:15	Lunch Break	
16	14:15	15:15	Test – 5 Heat Load Test at 100% load with adjacent CRAC stop (CRAC Unit 4 & 5)	

Item	From	To	Tasks	Remarks
17	15:15	16:15	Test – 6 Heat Load Test at 100% load with adjacent CRAC stop (CRAC Units 10 & 11)	
18	16:15	17:15	Test – 7 Heat Load Test at 100% load at Fire Trip shutting down all CRAC in data Centre	
19	17:15	18:15	Test 8 – Test UPS Room CRAC Failure at 100% I.T. Load	
			Next Working Day	
20	9:00	9:15	Registration for duty reporting	
21	9:15	9:30	Technicians set up test condition.	
22	9:30	9:45	Roll call / Pre-Test briefing	
23	9:45	10:00	Walkie Talkie Radio check	
24	10:00	10:05	Instrumentation check	
25	10:05	10:15	Personnel location check	
26	10:15	10:30	Test commencement announcement	
27	10:30	11:00	Normal Condition data centre operation check	

28	11:00	12:00	Test – 9 Heat Load Test at 100% load with Battery Room FCU Failure	
29	12:00	13:00	Test – 10 Heat Load Test at 100% Load with riser (path A) failure	
30	13:00	14:00	**Lunch Break**	
31	14:00	15:00	Test – 11 Heat Load Test at 100% Load with riser (path B) failure	
32	15:00	16:00	Test – 12 CRAC Communication Failure at 100% Load	
33	16:00	17:00	De-briefing upon Heat Load Test completion Submit all Heat Load Test record	

APPENDIX 2

Data Logger Layout

www.ingramcontent.com/pod-product-compliance
Lightning Source LLC
Chambersburg PA
CBHW030742180526
45163CB00003B/892